FREE BOOKS + MATH LESSONS:
www.Coursenvy.com/Math

6th Grade Math Workbook
Published by Coursenvy®
Los Angeles, California
info@coursenvy.com

Table of Contents

Table of Contents

1.1 - Ratios

The *ratio* is a comparison of the relative sizes of two or more values.

EXAMPLE: A circular box can hold 12 small balls. Four balls are green, and the remaining balls are blue. What is the ratio of green balls to blue balls?
Solution:

Find the number of blue balls by subtracting the number of green balls from the total number of balls. 12 total balls - 4 green balls = 8 blue balls

Hence, the ratio of the green balls to blue balls is 4/8. Express the ratio in its simplest form, ½ or 1:2 in colon notation.

The ratio of green balls to blue balls is 1:2.

Answer: **1:2**

To reduce ratio A:B easily, list the prime factors of A and B. Cross out the similar factors in both A and B. Then, multiply the remaining factors.

EXAMPLE: Reduce the ratio 25:45 to the simplest form.

Solution:

$$25 : 45 = \frac{25}{45} = \frac{5 \times 5}{5 \times 3 \times 3} = \frac{5}{3 \times 3} = \frac{5}{9} = 5 : 9$$

Answer: **5:9**

EXAMPLE: Liza has three dogs and two cats. What is the ratio of cats to dogs?
Solution:

The ratio of cats to dogs equals the number of cats divided by the number of dogs = ⅔.

Answer: **2:3**

1.1 - Ratios

1) There are 20 pens in a cup. Two pens are red, 15 pens are black, and the remaining pens are blue. What is the ratio of red pens to blue pens?

2) A circular box can hold 62 small balls. Twenty balls are yellow, 12 balls are green, and the remaining balls are blue. What is the ratio of green balls to blue balls?

3) There are 90 desks in a class. Twenty-four desks are black, and 54 desks are blue. The remaining desks are green. What is the ratio of blue desks to green desks?

4) In an assignment week, each student can solve ten math problems on Algebra, 15 math problems on Arithmetic, and 12 math problems on Geometry. What is the ratio of the Arithmetic to Geometry problems that each student can solve?

5) The angles A and B of a triangle ABC are 30° and 45°, respectively. If we denote the triangle's remaining angle as C, what is the ratio of angle B to angle C?

1.1 - Ratios

6) An alloy that weighs 125 g is a mixture of iron, zinc, copper, and nickel. The amount of iron is 47 g, zinc is 32 g, and copper is 23 g. What is the ratio of zinc to nickel?

7) There were 125 comic books, 200 detective books, and 1200 other books in a local library. The library authority decided to arrange the comic books on five shelves. What is the ratio of comic books to shelves?

8) There were 125 comic books, 200 detective books, and 1200 other books in a local library. The library authority decided to arrange the detective books on five shelves. What is the ratio of detective books to shelves?

9) In an orientation program, 125 seats were allotted for female students, 140 seats for male students, and 70 seats for school officials. What is the ratio of the number of seats for male students to female students' number of seats?

10) In a fruit basket, there are 24 total fruits. There are seven apples, five bananas, and an equal number of oranges and avocados. What is the ratio of apples to avocados?

1.1 - Ratios

11) The width and the length of a rectangle is 24 m and 72 m, respectively. What is the ratio of the length to the breadth of the rectangle in its lowest term?	12) There are two squares of sides 3 and 7, respectively. What is the ratio of their areas in its lowest term?
13) Let angle A and angle B be the complementary angles. If angle A equals 60°, what is the ratio of angle B to angle A?	14) Let angle A and angle B be the supplementary angles. If angle A equals 60°, what is the ratio of angle B to angle A?
15) A special dessert item that weighs 500 g requires powdered milk and sugar to be prepared. The chef adds 125 g of sugar. What is the reduced ratio of milk to sugar to the lowest term?	16) The weights of Devi and Rosa are 48 kg and 56 kg, respectively. What is the reduced ratio of their weights to the lowest term?
17) The heights of Devi and Rosa are 160 cm and 165 cm, respectively. What is the reduced ratio of their heights to the lowest term?	18) Reduce the ratio of 57:36 to the lowest term.
19) The numbers of male students and female students in a class are 300 and 270, respectively. What is the reduced ratio of male to female students to the lowest term?	20) Reduce the ratio of 92:36 to the lowest term.

1.1 - Ratios

21) Reduce the ratio 8:12 to the simplest form.	22) Reduce the ratio 13:65 to the simplest form.
23) There are five classrooms for 150 students. Reduce the ratio of classrooms to students to the simplest form.	24) Jacob has 15 books and 12 pencils. What is the reduced ratio of books to pencils?
25) Reduce the ratio of 75:105 to the simplest form.	26) Reduce the ratio of 45:85 to the simplest form.
27) Reduce the ratio 16:26 to the simplest form.	28) Let angle A and angle B be the supplementary angles. If angle A equals 30°, what is the ratio of angle B to angle A?
29) There are two squares of sides 5 and 9, respectively. Reduce the ratio of the areas of the smaller to larger square to the simplest form.	30) There are two squares of sides 2 and 3, respectively. Reduce the ratio of the areas of the smaller to larger square to the lowest term.

1.2 - Unit Rates

A *unit rate* is a ratio where one of the terms is 1.

EXAMPLE: What is the unit rate of 120 students for every six classrooms?

Solution: Here, the number of students for each classroom is simply the unit rate.

For every six classrooms, there are 120 students.

Hence, for one classroom, there are $\frac{120}{6}$ or 20 students.

The unit rate is 20 students per classroom.

Answer: **20 students per classroom**

EXAMPLE: Liana bought 90 candies for 45 kids. Find the ratio as a rate.

Solution:

Forty-five kids get 90 candies.

Hence, a kid gets $\frac{90}{45}$ or 2 candies.

The ratio as a rate is two candies per kid.

Answer: **2 candies per kid**

EXAMPLE: There are 255 books placed on 17 shelves equally. What is the unit ratio?

Solution:

For 17 shelves, there are 255 books placed equally.

Hence, there are $\frac{255}{17}$ or 15 books per shelf.

Answer: **15 books per shelf**

1.2 - Unit Rates

31) What is the unit rate of 180 students for every six classrooms?	32) The rate of students to buses is 60 to 3; express the ratio as a unit rate.
33) If the ratio of kids to ice creams is 50 to 150, what is the unit ratio?	34) Find the unit ratio of 575 fruits for every 23 baskets.
35) A photocopy machine can make 104 copies in 8 minutes. Find the ratio as a unit rate.	36) John earns $160 for every 5 hours. What is the unit rate?
37) If there are 318 pencils for 159 students, what is the ratio as a unit rate?	38) A solution has 60 grams of water and 20 grams of vinegar. Find the unit ratio.
39) A librarian arranged 675 books into 27 shelves equally. What is the unit ratio?	40) Laura bought 270 candies for 45 kids. Find the ratio as a unit rate.

1.2 - Unit Rates

41) Alexandra bought seven books, each at the same price, for $91 total. Find the unit ratio.	42) Each employee will work 360 hours in 60 days. What is the ratio as a rate?
43) Every 2 feet of wall decoration costs $3. If the wall is 48 feet in height, what is the cost in total?	44) Sally can make 50 cookies in 45 minutes. She needs to make 150 cookies altogether. How long will it take her to make all the cookies?
45) Six students can form two groups. If the school authority wants to make exactly 18 groups, what proportion is required to determine the number of students (x)?	46) Liz wants to buy some colored pencils. Patrick's Stationery sells eight pencils for $4. Fancy Goods sells 12 pencils for $5. Where should Liz buy pencils to minimize her expense?
47) A salesperson can sell at least two shirts in 25 minutes. The company has a direction to sell at least 30 shirts in his 8-hour shift. Will the salesperson fulfill the company's requirements?	48) An amount of $3 is needed to wrap the five square feet floor. If the floor is 1,200 square feet in total, what will be the wrapping cost?
49) A worker can work only 9 hours a day. In that way, how long will it take to make a building that requires 558 hours to build?	50) John can make a painting in 45 minutes. He needs to make four paintings together. How many hours will it take him to make all the paintings?

1.2 - Unit Rates

51) Monica needs to solve 16 assignment problems in two hours. How many problems should she solve in an hour?	52) David walks 12 miles every 4 hours. What is his speed rate?
53) A baker needs to buy bananas for the cake. The banana costs $3 for 18 pieces at Shop A, $4 for 28 at Shop B, and $7 for 35 bananas at Shop C. From which shop would the baker get more bananas per dollar?	54) Laurel paid $52 for 13 copies of a nonfiction book. What is the price of one book?
55) Zach can run 14 miles for 4 hours. What is his speed rate?	56) If five paintings cost $85, what is the price per painting?
57) Jennifer can make 20 cupcakes in 30 minutes. She needs to make 80 cupcakes all together. How long will it take her to make all the cupcakes?	58) Zachary put 285 books on 19 shelves. What is the unit ratio?
59) A printer can print 90 pages in 18 minutes. Find the ratio as a unit rate.	60) What is the unit rate of 72 students to 24 benches?

1.3A - Equivalent Ratio Tables

Equivalent ratios are two ratios that have the same values. These can be obtained by multiplying or dividing both numerator and denominator by the same number.

EXAMPLE: Find the value of x, for which 20:x = 5:27.

Solution: Cross multiply the two known values and divide the product by the remaining value. 20:x = 5:27

$$\frac{20}{x} = \frac{5}{27} \text{ so, } x = \frac{20 \times 27}{5} = 108$$

Answer: **108**

EXAMPLE: For what number will the table give equivalent ratios?

5	10
	52

Cross multiply the two known values and divide the product by the remaining value.

$$\frac{5}{10} = \frac{1}{2}$$

So, $\frac{x}{52} = \frac{1}{2}$ $x = \frac{52}{2} = 26$

Answer: **26**

EXAMPLE: Is 3:4 equivalent to 12:16?

Remember, we can create an equivalent ratio by multiplying or dividing both parts of the ratio by the same number.

Consider, $12 : 16 = \frac{12}{16} = \frac{(3 \times 4)}{(4 \times 4)} = \frac{3}{4}$

Hence, 3:4 is equivalent to 12:16

Answer: **Yes**

1.3A - Equivalent Ratio Tables

61) What is the number for the ratio ___: 12 that will make an equivalent ratio to 1:3?

62) What is the number for the ratio ___: 32 that will make an equivalent ratio to 3:4?

63) What is the number for the ratio 20: ___ that will make an equivalent ratio to 5:27?

64) Shop A: $27 for six burgers.
   Shop B: $36 for eight burgers.

Are these equivalent ratios?

65) 39 cupcakes: 13 kids
   65 cupcakes: 26 kids

Are these equivalent ratios?

1.3A - Equivalent Ratio Tables

66) For what number does the table give equivalent ratios?

45	90
	52

67) For what number does the table give equivalent ratios?

9	4
27	

68) Find the numbers for equivalent ratios?

6	2
27	
	12

69) Lisa works 3 hours for $45. Using equivalent ratios, find the following numbers.

3 hours	$45
8 hours	
5 hours	

70) Using equivalent ratios, find the following numbers.

3 eggs	12 cupcakes
7 eggs	
15 eggs	

1.3B - Unit Rate and Unit Price Word Problems

EXAMPLE: Two sides of a rectangle measures 3 and 5 units. What is the ratio of the perimeter to the area of the rectangle?

Given: length (L) = 5 units, width (W) = 3 units
Asked: the ratio of perimeter to the area of the rectangle
Operation: addition, multiplication

Solution: The formula for finding the perimeter of the rectangle is P = 2 (L + W).

Hence, the perimeter of the rectangle is 2 (5 + 3) = 16. The formula for finding the area of the rectangle is A = L × W. Hence, the area of the rectangle is 5 × 3 = 15. Now, the ratio of the perimeter to the area is 16/15.

Answer: **16:15**

EXAMPLE: A fruit shop prices $3 for five bananas. How many bananas can be bought for $12?

Given: Five bananas for $3 = 5/3 per banana. $3 for five bananas, $12 for x bananas.
Asked: number of bananas can be bought for $12.
Operation: division, multiplication

Solution:
$$\frac{5}{3} \times 12 = 20 \text{ bananas}$$

Answer: **20 bananas**

EXAMPLE: A pencil box can hold 15 pencils. If there are 90 pencils, how many boxes are required?

Given: 15 pencils in 1 box, 90 pencils in x boxes
Asked: number of boxes required to hold 90 pencils
Operation: division, multiplication

Solution:
$$\frac{1}{15} \times 90 = 6 \text{ boxes}$$

Answer: **6 boxes**

1.3B - Unit Rate and Unit Price Word Problems

71) Peter can read 12 pages in 1 hour. How much time will he need to read a book with 576 pages?	72) Lily traveled 3 miles in 1 hour. How many miles did she travel per minute?
73) Where should Sandy buy to minimize her expense? <table><tr><td>Shop A</td><td>$2</td><td>16 candies</td></tr><tr><td>Shop B</td><td>$5</td><td>35 candies</td></tr></table>	74) Which product has the best price? <table><tr><td>Product A</td><td>$35</td><td>90 kg</td></tr><tr><td>Product B</td><td>$40</td><td>124 kg</td></tr></table>
75) What is the unit rate of 180 passengers for every six buses?	76) Lexi bought a bundle of nine books for $270. What is the unit price?
77) Lola had to walk 2 miles in 1 hour. How long did she walk for each mile?	78) Two sides of a rectangle measure 8 and 12 units. What is the ratio of the perimeter to the area of the rectangle?
79) Jill works 5 hours for $60. How much does she earn hourly?	80) Deb works 4 hours for $50. How much does he earn hourly?

1.3B - Unit Rate and Unit Price Word Problems

81) In a factory, a machine takes 5 minutes to pack 1000 chocolate bars. How much time will it take to pack 98,000 chocolate bars?	82) The subscription fees for two different websites are in the table below. What is the best deal? A (2 months) — $80 — unlimited C (4 months) — $130 — unlimited
83) Teddy needs three bananas to bake 15 pancakes. How many bananas does he need to bake 35 pancakes?	84) Rubi solves 20 problems in 1.5 hours. How many problems does she solve in 4.5 hours?
85) The small apartment rent for eight weeks costs $104. How much does the rent cost for three weeks?	86) The prices of shirts in different shops are shown in the table below. Which shop is offering the best rate? A — $130 — 15 shirts B — $100 — 11 shirts
87) Eric buys three umbrellas for $10.50. How much does he need to pay for five umbrellas?	88) Ninety bricks are needed to build a wall. How many bricks are needed to make four walls?
89) Grandma wants to give eight candies to two kids equally. If there are 18 kids, what proportion is required to determine the number of candies?	90) A pencil box can hold 15 pencils. If there are 90 pencils, what proportion is required to determine the number of pencil boxes?

1.3C - Percentages

A *percentage* is a number that represents a fraction of 100. The word "*percentage*" comes from the Latin phrase "*per centum*," which means "*by the hundred.*"

EXAMPLE: Convert the decimal to fraction. Convert 0.55 to an equivalent fraction.

Solution: Steps to Convert Decimal to Fraction
1. Write the decimal number as the numerator and a power of 10 as the denominator.
2. Find the greatest common factor (GCF) of both numerator and denominator.
3. Reduce the fraction by dividing the numerator and denominator by the GCF.

$$0.55 = \frac{55}{100} = \frac{55 \div 5}{100 \div 5} = \frac{11}{20}$$

Answer: $\frac{11}{20}$

EXAMPLE: Convert the percent to fraction. Convert 45% to an equivalent fraction.

Solution: Steps to Convert Percent to Fraction
1. Write the percent number as the numerator and 100 as the denominator.
2. Find the greatest common factor (GCF) of both numerator and denominator.
3. Reduce the fraction by dividing the numerator and denominator by the GCF.

$$45\% = \frac{45}{100} = \frac{45 \div 5}{100 \div 5} = \frac{9}{20}$$

Answer: $\frac{9}{20}$

EXAMPLE: What is 50% of 160?

Solution: To calculate the percentage of a number, change the rate into a decimal by dividing it by 100. Then, multiply it by the base number.

Calculation Trick: x% of y = y% of x → Example: 5% of 40 is the same as 40% of 5. 40% of 5 is 2. Hence, 5% of 40 is 2.

$$50\% \text{ of } 160 = \frac{50}{100} \times 160 = 0.5 \times 160 = 80$$

Answer: **80**

1.3C - Percentages

91) Write the decimal 0.33 as a fraction.	92) Determine the fraction when the decimal is 5.80.
93) Determine the fraction to the reduced form when the decimal is 5.80.	94) Express the decimal 0.9950 as a reduced fraction.
95) Write the decimal 0.0203 as a fraction.	96) Change the decimal 0.5 to a percent.
97) Change the decimal 0.07 to a percent.	98) Write the decimal 0.0203 as a percent.
99) Express the decimal 1.2 as a percent.	100) Change the decimal 0.025 to a percent.

1.3C - Percentages

101) Change 25% to a fraction.	102) Change 25% to a reduced fraction.
103) Convert 1.32% to a reduced fraction.	104) Convert 78% to a reduced fraction.
105) Convert 0.5% to a reduced fraction.	106) Change 30% to a decimal.
107) Change 1.56% to a decimal.	108) Convert 3/2 to a decimal.
109) Express 3/6 into a percentage.	110) Express 2/5 into a percentage.

1.3C - Percentages

111) What is 30% of 160?	112) What is 21% of 450?
113) What is the price of a book with a 20% discount if the actual price is $20?	114) Micheal bought a book with a 30% discount. The actual price was $45. How much was the discount?
115) In a mixture of 4 kg of sugar and flour, 25% is sugar. Determine the quantity of sugar in the mixture.	116) In a mixture of 4 kg of sugar and flour, 75% is flour. Determine the quantity of flour in the mixture.
117) Garden soil is a mixture of soil and 45% fertilizer. For 400 kg of garden soil, determine the quantity of fertilizer.	118) In a library, comic books are 32% of the total collection. There are a total of 16,000 books in that library. How many are comic books?
119) Rosie got 89% marks in her exam. The exam had a total of 900 marks. What were her obtained marks?	120) David earns $1500 per month. He spends 25% on food. How much does he spend on food?

1.3C - Percentages

121) In a shop, there are 560 fruits, of which 30% are mangoes. How many fruits are mangoes?	122) 500 students are studying in a school. On Sunday, the percentage of the students was 95%. How many students were absent on Sunday?
123) There are 500 students in a school. On Sunday, the percentage of the students was 95%. How many students were present on Sunday?	124) Monica and Maria bought 70 berries and 90 berries, respectively. What percent of the total berries did Monica buy?
125) What percent of $200 is $100?	126) Jami paid $120 for some greeting cards. She had to pay a 10% tax. What was the total amount that she had to pay?
127) What percent of $300 is $120?	128) In a locality of among 500 people, 240 people are jobholders. What percent of total people are jobholders?
129) A fruit basket contains some fruits. Two fruits out of every 16 fruits are rotten. What is the percentage of rotten fruits?	130) What percent of 120 equals 12?

1.3D - Ratios and Units of Measurement

EXAMPLE: One yard equals 3 feet. How many feet are in 23 yards?

Solution:

It's known that 1 yard = 3 feet.

Hence, 23 yards = 23 x 3 feet = 69 feet

Answer: **69 feet**

EXAMPLE: How many 4-ounce containers does Ronald need to have 12 gallons in total?

Solution:

It is known that 128 ounces = 1 gallon.

1 container = 4 ounces
(1/4) x 128 ounces = 32 containers

Hence, 1 gallon = 32 containers
32 containers per gallon x 12 gallons = 384 containers

Answer: **384 containers**

EXAMPLE. Two acres of land produce 60 bushels of corn. Three acres of land produce 90 bushels of corn. How many bushels of corn can be produced on four acres of land?

Solution:
Let the number of bushels of corn that can be produced on 4 acres of land = x.

Then, x : 4 = 90 : 3 = 60 : 2 = 30 : 1

$$\frac{x}{4} = \frac{30}{1}$$

x = 30 × 4

x = 120

Answer : **120 bushels of corn**

1.3D - Ratios and Units of Measurement

131) One yard equals 3 feet. How many feet are in 16 yards?	132) Some ratios for the yard to inches conversion are 2:72 and 3:108. How many inches are in 5 yards?
133) There are 12 inches in 1 foot. Three feet equals 1 yard. How many inches are in 5 yards?	134) There are 5,280 feet in a mile. How many yards are in 2 miles? [1 yard = 3 feet]
135) There are 5,280 feet in a mile. How many inches are in 1/2 miles? [12 inches = 1 foot]	136) If there are 12 inches in 1 foot, how many feet are 3 inches?
137) If there are 36 inches in 1 foot, how many inches are 7 feet?	138) Convert 4 feet to yards. [1 yard = 3 feet]
139) Express 52 inches into feet. [1 feet = 12 inches]	140) What is 9 feet in yards? [1 yard = 3 feet]

1.3D - Ratios and Units of Measurement

141) One minute equals 60 seconds. How many seconds are 4.5 minutes?	142) 150 seconds = _____ minutes
143) 5 minutes and 30 seconds = _____ seconds	144) 5 minutes and 25 seconds = _____ minutes
145) Robert can swim for 45 minutes. How many hours can he swim?	146) There are 52 weeks in 1 year. How many weeks are in ¼ years?
147) How many cups of 4 ounces does Ted need to have 2 gallons? [128 ounces = 1 gallon]	148) How many centimeters are in 2.4 meters?
149) There are 4 quarts in 1 gallon. How many quarts are in ¾ gallon?	150) Tiffany bought 4 kilograms and 500 grams of peanuts. How many peanuts (in grams) did she buy?

1.3D - Ratios and Units of Measurement

151) Convert 3.2 pounds into ounces. [1 lb = 16 oz]	152) Which is more, A. 7 lb 8 oz or B. 124 oz?
153) A 2-feet and 3-inch rod was stuck to a 1-foot rod. What is the length of the later rod in inches?	154) Which is more, A. 3 lb 3 oz or B. 33 oz?
155) 4,380 milliliters =______ deciliters	156) The price of a 1-gallon chemical bottle is \$3.5. What is the price per pint? [1 gallon = 8 pints]
157) Shop A offers 2 gallons and 3 pints of liquid for \$4.8. Shop B offers 2.5 gallons of liquid for \$4.8. Which shop is offering the better deal?	158) 2 gallons and 3 pints + 2.5 gallons = ______ gallons
159) 2.5 gallons - 2 gallons and 3 pints = ______ pints	160) (1 lb 5 oz) $\times$ 2 = ___ oz

2.1 - Divide Fractions

EXAMPLE: What is the product of $\frac{3}{5} \times \frac{2}{7}$?

Solution: To multiply two fractions, multiply across the numerators and across the denominators. Simplify the fraction, if necessary.

$$\frac{3}{5} \times \frac{2}{7} = \frac{3 \times 2}{5 \times 7} = \frac{6}{35}$$

Answer: $\frac{6}{35}$

EXAMPLE: $\frac{4}{7} \div \frac{1}{2} = $?

Solution: To divide two fractions, invert the divisor, then multiply across the numerators and across the denominators. Simplify the fraction, if necessary.

Step 1: Invert the divisor.
$$\frac{4}{7} \div \frac{1}{2} = \frac{4}{7} \times \frac{2}{1}$$

Step 2: After inverting the divisor, multiply the fractions.
$$\frac{4}{7} \times \frac{2}{1} = \frac{4 \times 2}{7 \times 1} = \frac{8}{7}$$

Answer: $\frac{8}{7}$

EXAMPLE: How many 1/5 cup servings are in 2 cups of yogurt?

Solution: If we divide two equally by $\frac{1}{5}$, then
$$2 \div \frac{1}{5} = 2 \times \frac{5}{1} = 2 \times 5 = 10$$

Thus, there are **10** cups of servings in 2 cups of yogurt.

Answer: **10 cups**

2.1 - Divide Fractions

161) What is the product of $\frac{5}{12} \times \frac{9}{10}$?	162) What is the product of $\frac{3}{19} \times \frac{38}{45}$?
163) What is the product of $\frac{7}{9} \times \frac{4}{21}$?	164) $\frac{5}{12} \times \frac{5}{10} = ?$
165) A pancake box is $\frac{2}{3}$ full. A customer comes in and buys $\frac{1}{3}$ of pancakes. How much of the box does the customer buy?	166) $\frac{2}{21} \times \frac{45}{3} = ?$
167) What is the reduced equivalent fraction for $\frac{15}{45}$?	168) $\frac{9}{28} \div \frac{18}{7} = ?$
169) How much banana will each person get if four people share 24 bananas equally?	170) How many 1/4 cup servings are in 2 cups of yogurt?

2.1 - Divide Fractions

171) How wide is a rectangular pond with a length of 3/4 meter and an area of 3/2 square meters?	172) A man gave 1/8 of his property to his wife, 1/2 to his son, and 1/4 to his daughter. The value of the remaining property is $60,000. What percent of the property remained after his gifting?
173) $\frac{5}{12} \div \frac{35}{24} = $?	174) $\frac{18}{7} \div \frac{36}{35} = $?
175) Sofia has $\frac{5}{6}$ of a tank of fuel in her car. She uses $\frac{1}{6}$ of a tank per day. How many days will her fuel last?	176) Baby pajamas require $\frac{1}{3}$ yard of fabric. How many pajamas can be made from 6 yards of fabric?
177) How many one-fourth pies are in 20 whole pies?	178) How many two-sixth are in 18 pizzas?
179) How many one-fourths are in 6 stems of broccoli?	180) How many half centimeters are in 20 centimeters?

2.1 - Divide Fractions

181) Divide: 2 ÷ (2/5)	182) Divide: 10 ÷ (5/6)
183) Divide: 6 ÷ (6/7)	184) Divide: 3 ÷ (1/6)
185) Divide: 5 ÷ (4/9)	186) Divide fractions. Reduce the quotient to its lowest term. (1/5) ÷ (6/5)
187) Divide fractions. Reduce the quotient to its lowest term. (7/12) ÷ (3/4)	188) Divide fractions. Reduce the quotient to its lowest term. (8/9) ÷ (16/9)
189) Divide fractions. Reduce the quotient to its lowest term. (⅕) ÷ (4/9)	190) The product of two fractions is 4/8. If one of the fractions is 3/4, what is the other fraction?

2.1 - Divide Fractions

191) The product of two fractions is 4/3. If one of the fractions is 2/4, what is the other fraction?	192) The product of two fractions is 1/9. If one of the fractions is 1/6, what is the other fraction?
193) Fill in the blank to make the equation true: $\frac{1}{7} \times \frac{2}{3} = \frac{1}{7} \div \_$	194) Fill in the blank to make the equation true: $\frac{9}{8} \times \frac{3}{4} = \_ \div \frac{4}{3}$
195) Fill in the blank to make the equation true: $\frac{13}{5} \times \frac{21}{23} = \frac{13}{5} \div \_$	196) Fill in the blank to make the equation true: $\frac{17}{24} \times \_ = \frac{17}{24} \div \frac{19}{7}$
197) Patel can walk 1/6 of a mile in an hour. If Patel is 1/5 of a mile away from his school, how long will it take Patel to reach the school?	198) Cyndi is painting walls. She has 1/5 of a liter of painting remaining. If each wall requires 1/10 of a liter of pain, how many walls can she paint?
199) David has $4\frac{2}{3}$ cakes in his home. He cut the cakes into pieces that are each $\frac{1}{4}$ of a whole cake. How many pieces of cakes does he have?	200) Susan needs to repair $12\frac{1}{2}$ inches of a wall. She can repair $2\frac{1}{2}$ inches of walls per week. How many weeks will it take to repair her wall?

2.1 - Divide Fractions

201) $\frac{1}{4}$ liter of paint can cover a $3\frac{3}{4}$ m^2 wall. How many m^2 can you paint with 1 liter?	202) $4\frac{1}{3}$ liters of paint can cover a 65 m^2 wall. How many m^2 can you paint with 3 liters?
203) A 4 m plastic pipe weighs $3\frac{1}{2}$ pounds. How many pounds does 8 m of the pipe weigh?	204) Abraham can walk 4.5 miles in 2 hours. How many miles does Abraham walk in 4 hours?
205) Nathalia makes 2.5 pounds of cake in 30 minutes. How many cakes does she make in 60 minutes?	206) Every month, a chef needs $3\frac{1}{4}$ pounds of rice. How many pounds of rice are needed in 4 months?
207) If you cut $\frac{13}{4}$ m of an electric plumb into $\frac{1}{4}$ m sections, how many pieces it will be?	208) Fernandez can swim ½ miles in 20 minutes. Sarah can swim ¾ miles in 20 minutes. Who swims faster?
209) How many ½ centimeters are in 4 centimeters?	210) Susan can walk 2/3 of a mile in an hour. If Susan is 1/5 of a mile away from his school, how long will it take Patel to reach the school?

2.2 - Multi-Digit Division

EXAMPLE: What is the remainder of 63,973 ÷ 100?

Solution:

```
          6 3 9
  1 0 0 ) 6 3 9 7 3
          6 0 0
          3 9 7
          3 0 0
          9 7 3
          9 0 0
            7 3
```

Answer: **73**

EXAMPLE: What is the quotient of 63,973 ÷ 100?
Solution:

```
          6 3 9
  1 0 0 ) 6 3 9 7 3
          6 0 0
          3 9 7
          3 0 0
          9 7 3
          9 0 0
            7 3
```

Answer: **639**

EXAMPLE: Timmy spent $130 in 5 weeks. He spent the same amount each week. How much did he spend each week?

Solution: We divide $130 by 5.

130 ÷ 5 = 26

He spent $26 each week.

Answer: **$26**

2.2 - Multi-Digit Division

211) What is the quotient of 1,550 ÷ 10?	212) What is the quotient of 648 ÷ 18?
213) What is the resulting quotient of 962 ÷ 26?	214) 4,968 ÷ 36 = ?
215) 2,436 ÷ 58 = ?	216) 3,920 ÷ 28 = ?
217) What is the quotient of 43,440 ÷ 12?	218) What is the remainder of 4,340 ÷ 13?
219) 314,940 ÷ 87 = ?	220) 101,460 ÷ 57= ?

2.2 - Multi-Digit Division

221) What is the remainder of 1,453 ÷ 33?	222) Calculate 88,040 ÷ 62.
223) What is the quotient of 650 ÷ 25?	224) A ninja spent $250 in 5 weeks. He spent the same amount each week. How much did he spend each week?
225) 420 apples will be divided among 42 children equally. How many apples does each child get?	226) There are 1,225 chocolates in a jar. If there are five children, how many chocolates does each child receive?
227) What is the resulting quotient of 2225 ÷ 25?	228) How many pens will each student get if twelve students share 144 pens equally between them?
229) 114 pieces of rods are needed to prepare a table. How many tables can be prepared by 1,368 rods?	230) Cyndi saves $50 each month. How many months does Cyndi need to save a total of $600?

2.2 - Multi-Digit Division

231) There are 250 households in a town, and they are willing to donate \$30,000 total for constructing a theme park. How much money does each household need to give to the fundraiser (equally)?	232) $50{,}048 \div 368 = ?$
233) Every 230 products are packed in a box. In order to pack 50,000 products, how many boxes will be necessary?	234) A shopkeeper has 10,000 berries to put into jars. It takes 100 berries to fill one jar. How many jars will the shopkeeper need to fill with berries?
235) Divide 8475 by 565.	236) $15{,}100 \div 755 = ?$
237) Divide 10,540 by 155.	238) $677 \div 15 = ?$ What is the quotient and the remainder?
239) $2412 \div 10 = ?$ Write your answer as a whole number and remainder.	240) $3265 \div 100 = ?$ What is the quotient and the remainder?

2.2 - Multi-Digit Division

241) $43,932 \div 523 = ?$ What is the quotient and the remainder?	242) $97,500 \div 186 = ?$ What is the quotient and the remainder?
243) $2,989 \div 91 = ?$ Write your answer as a whole number and remainder.	244) $3,011 \div 74 = ?$ What is the quotient and the remainder?
245) $1,934 \div 82 = ?$ Write your answer as a whole number and remainder.	246) $2,213 \div 51 = ?$ Write your answer as a whole number and remainder.
247) $3,212 \div 32 = ?$ Write your answer as a whole number and remainder.	248) $1,950 \div 50 = ?$ Write your answer as a whole number and remainder.
249) $45,340 \div 145 = ?$ Write your answer as a whole number and remainder.	250) $34,220 \div 236 = ?$ Write your answer as a whole number and remainder.

2.2 - Multi-Digit Division

251) How many ¼ liters are in 4 liters?	252) There are 720 grams of flour in a jar. If 180 grams of rice are eaten every day, when will the flour be consumed?
253) Pam walks 3 miles in an hour. How many miles can she walk in 3 hours?	254) David can write a paper in 20 minutes. How many papers can David write in 60 minutes?
255) There are 6,780 apples in a farmhouse. A farmer needs to bag all the apples. If a bag contains 20 apples, how many bags does the farmer need?	256) A factory can produce 430 products in 7 days. How many products can it produce in 21 days?
257) Bobby bought ten pencils at $50. How much will he pay if he buys 30 pencils?	258) If 5 pounds of beef costs $20, how many pounds of beef can be bought for $6000?
259) A truck can travel 300 miles on 12 liters of gas. How many liters of gas will be necessary to travel 100 miles?	260) Susie bought 24 notebooks at $120. How much will he pay if he buys 30 notebooks?

2.3 - Decimals

A *decimal* is a number whose whole number and the fractional part is separated by a decimal point. The digits after the decimal points denote a value less than one.

EXAMPLE: 1.1 + 2.5 = ?

Solution:
To add decimals, write down the numbers, one under another, ensuring that decimal points are lined up. Then, add using column addition. Put the decimal point in the answer.

```
      1.1
  +   2.5
      3.6
```

Answer: **3.6**

EXAMPLE: What is 2.9 multiplied by 100?

Solution: Multiply decimals the same way as multiplying the whole numbers. Count the number of decimal places in each factor. Put the same number of decimal places in the product.

In number 2.9, there is one decimal place. In the number 100, there is no decimal. Now, multiply the two factors.

$2.9 \times 100 = 290.0$

We move decimal at one decimal place. This is a very important step to consider when multiplying decimal numbers.

Answer: **290**

EXAMPLE:
A fruit basket weighs 6.4 lbs. What is the weight of 10 fruit baskets?

Solution:
We want to multiply 6.4 by 10. Thus, we get
$6.4 \times 10 = 64$.

The total weight of 10 fruit baskets is 64 lbs.

Answer: **64 lbs**

2.3 - Decimals

261) $0.1 + 0.5 = ?$	262) $1.1 + 2.5 = ?$
263) $23.50 + 32.12 = ?$	264) $748.114 + 125.168 = ?$
265) 21 hundredths + 9 hundredths = ?	266) 37 hundredths + 17 hundredths =
267) $0.9 - 0.3 = ?$	268) $8.4 - 5.2 = ?$
269) $94.79 - 34.09 = ?$	270) $816.321 - 413.321 = ?$

2.3 - Decimals

271) 990.103 – 770.100 = ?	272) 8 tenths – 4 tenths =?
273) 33 thousandths – 27 thousandths =?	274) 3.4 × 5 = ?
275) 9.82 × 10 = ?	276) 8.41 × 14 = ?
277) 6 × 5 ones =	278) 8 × 3 tenths = ?
279) b × (c+d) = ?	280) 8 × (3.4+1.2) = ?

2.3 - Decimals

281) $19.82 \times 10.4 = ?$	282) $14.47 \times 23.72 = ?$
283) $7.103 \times 2.4 = ?$	284) $0.05 \times 100 = ?$
285) What is 0.7 multiplied by 100?	286) What is 2.9 multiplied by 100?
287) What is 4.4 multiplied by 10?	288) $87.18 \div 13 = ?$
289) $65.36 \div 19 = ?$	290) $95.39 \div 24 = ?$

2.3 - Decimals

291) 57.81 ÷ 6 = ?	292) 17.38 ÷ 3.2 =?
293) 83.14 ÷ 9.3 = ?	294) 24 tenths ÷ 4 tenths = ?
295) 22.4 ÷ 10 =?	296) 295.6 ÷ 100 =?
297) There are 4 cups of sugar, and each cup weighs 0.3 pounds. How many pounds do these 4 cups of sugar weigh?	298) You have $39.56. If you give $23.98 to your friend, how much money do you have now?
299) Sam earned $420.50 each week. If he worked four weeks consistently, how much money did Sam earn?	300) A packet contains 0.5 liters of milk. If we have 45 packets, how many liters of milk will there be?

2.3 - Decimals

301) Jack walked 6.5 miles on the first day, 3.2 miles on the second day, 3.5 miles on the third day, and 5.5 miles on the fourth day. How many miles did Jack walk in four days?	302) Emma bought a new pen costing $6.5. She gave $10 to the shopkeeper. How much change should she receive back?
303) Becky spends $9.5 for her lunch each day. How much money does she spend on her lunch in 30 days?	304) A fruit basket weighs 6.4 lbs. What is the weight of 10 fruit baskets?
305) Isabella wanted to buy ice cream. She had $9.8. The ice cream price was $10.5. How much money did she still need?	306) The cost of a computer is $250.5. What is the price of five computers?
307) Emily purchased 20 notebooks for $120.5. What is the cost of each notebook if each notebook cost equally?	308) At a fast food restaurant, the total bill was $27.95 for seven people. How much money should each person pay to split the bill equally?
309) A truck traveled 45 miles in 6 hours. How much distance did truck travel per hour?	310) A motorbike travels 0.01 mile per minute. How many miles will it travel in 8 minutes?

2.4 - Greatest Common Factor and Least Common Multiple

EXAMPLE:

What are the factors of 20?

Solution: The numbers that can divide 20 without a remainder are called *factors* of 20. The numbers 2, 4, 5, 10, and 20 can divide 20 without a remainder. Thus, the factors of 20 are 2, 4, 5, 10, and 20.

Answer: **2, 4, 5, 10, and 20.**

EXAMPLE:
Determine the LCM of the following numbers: 22, 44, 88, 176.

Solution: *LCM* stands for *least common multiple.* To determine the LCM, list enough multiples of each given number, then select the least common multiple.

Multiples of 22: 22, 44, 66, 88, 110, 132, 154, **176**
Multiples of 44:   44,   88,   132,   **176**, 220
Multiples of 176:   **176**, 352

The number 176 is the least number that is common among these multiples.

Answer: **176**

EXAMPLE: Determine the GCF of the following numbers: 24, 36, 48.

Solution: The *greatest common factor (GCF)* of numbers is the largest factor that numbers have in common. To determine the GCF, list all the possible factors of each given number, then select the largest common factor.

Factors of 24: 2, 3, 4, 6, 8, **12**, 24
Factors of 36: 2, 3, 4, 6, 9, **12**, 18, 36
Factors of 48: 2, 3, 4, 6, 8, **12**, 16, 24, 48

The greatest common factor of 24, 36, and 48 is 12.

Answer: **12**

2.4 - Greatest Common Factor and Least Common Multiple

311) What are the factors of 24?

312) Write the factors of 64.

313) What is the greatest common factor of 42 and 126?

314) What is the greatest common factor of 36 and 144?

315) What is the greatest common factor of 120 and 320?

2.4 - Greatest Common Factor and Least Common Multiple

316) Determine the LCM of the following numbers: 22, 44, 88, 176.

317) Determine the GCF of the following numbers: 144, 240, 612.

318) The LCM of two numbers is 24, and their GCF is 4. If one of the numbers is 12, find the other.

319) The LCM of two numbers is 144, and their GCF is 1. If one of the numbers is 9, what is the other one?

320) The LCM of two numbers is 60, and their GCF is 2. If one of the numbers is 10, what is the other one?

2.4 - Greatest Common Factor and Least Common Multiple

321) Determine the GCF of the following numbers: 240 and 480.

322) Find the least common multiple of 8 and 16.

323) Determine the GCF of the following numbers: 120, 240, 480.

324) What is the greatest number that divides 100 and 184 with both remainder of 4?

325) There are three bells of different colors. The red bell rings every 8 minutes, the yellow bell rings every 10 minutes, and the green bell rings every 20 minutes. If they ring together at 6 P.M., find the time when all ring together next time.

2.4 - Greatest Common Factor and Least Common Multiple

326) Mia, Tina, Luna, and Eve run a distance of 4 miles, 6 miles, 6 miles, and 8 miles, respectively, from a point after a definite interval of time. What is the minimum distance traveled by all of them when they meet together?

327) The length of a tin-sheet and an aluminum-sheet is 442 cm and 1,082 cm. What is the LCM for the two lengths?

328) Cindy walks 4 miles in an hour. Amanda walks 3 miles in 30 minutes. Who walks faster?

329) There are more than ten students in a class. The teacher wants to distribute 12 bananas, 48 peaches, and 96 candies to these students equally without leaving any remaining. To how many students can the teacher distribute the fruits and candies?

330) There are more than 20 students in a class. The teacher wants to distribute 16 apples, 48 oranges, and 192 chocolates to these students equally without leaving any remaining. To how many students can the teacher distribute the fruits and chocolates?

2.5 - Positive & Negative Integers

EXAMPLE: What are the integers between -2 and 6?

Solution:
An *integer* is a whole number that can be positive, negative, or zero.
Positive numbers are all the whole numbers greater than zero.
Negative numbers are all the whole numbers less than zero.

Here we write down all the whole numbers, including zero.

Thus, we write numbers between -2 and 6.

Answer: **-2, -1, 0, 1, 2, 3, 4, 5, 6**

EXAMPLE: - 6 + 4 = ?

Solution:
To add integers with unlike signs, subtract the absolute values, and prefix the sign of the higher absolute value.

To add integers with like signs, add the absolute values, and prefix the common sign.

Thus, - 6 + 4 = - 2.
- 6 and 4 have unlike signs. The difference of their absolute values is 2. The sign of the higher absolute value is negative.

Answer: **- 2**

EXAMPLE: (-30) - 7 = ?

Solution:
To subtract integers, keep the first number (known as *minuend*). Change the operation from subtraction to addition. Get the opposite of the second number (known as *subtrahend*). Proceed to regular addition of integers.

Thus, (-30) - 7 = (-30) + (-7) = - 37. (Keep - 30. Change subtraction to addition. The opposite of 7 is - 7. The expression now is (-30) + (-7). -30 and -7 has like signs. Hence, add their absolute values that is 37. Prefix the common sign that is negative. Therefore, the answer is -37.

Answer: **-37**

2.5 - Positive & Negative Integers

331) What is the starting point on horizontal and vertical number lines?	332) On a horizontal number line, the numbers above 0 are __________ numbers.
333) On a horizontal number line, the numbers below 0 are __________ numbers.	334) On a vertical number line, the numbers below 0 are __________ numbers.
335) On a vertical number line, the numbers above 0 are __________ numbers.	336) Write - 5, 7, 8, - 3, -1, 2, 1, 9 in order.
337) How many integers are between -3 and 4?	338) How many integers are between -9 and 1?
339) What are the integers between -4 and 4?	340) What are the integers between -7 and 15?

2.5 - Positive & Negative Integers

341) What are the integers between -10 and 5?	342) 11 + (-7) = ?
343) 1 + (-6) = ?	344) (-30) + (-7) = ?
345) 23 + (-10) = ?	346) 0 + (-1) = ?
347) 13 + (-13) = ?	348) - 12 - 7 = ?
349) 19 - 24 = ?	350) 110 - 113 = ?

2.6A - Opposite Numbers

EXAMPLE:
What is the opposite of 2?

Solution:
The *opposite number* of a positive number is the number that is less than 0 of a number line. Thus, the opposite of 2 is -2.

Answer: **-2**

EXAMPLE: Locate the opposite of 5/2 on the number line.

A *number line* is a representation of ordering real numbers. A *number line* is commonly horizontal, has points, and equally spaced that corresponds to each whole number (either positive, negative, or zero).

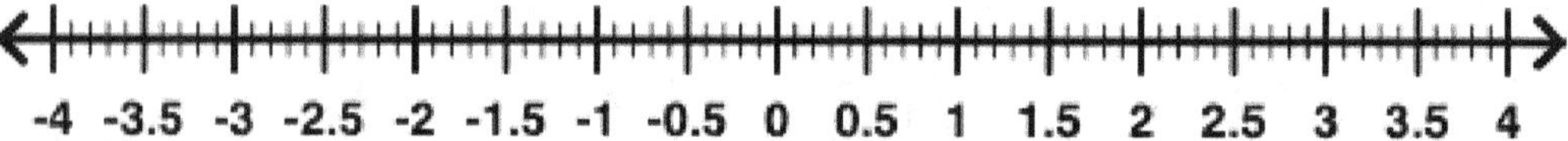

Solution: 5/2 can be written as 2.5. The opposite of 2.5 is -2.5, which is on the left side of the number 0 on the number line.

Answer: **-2.5**

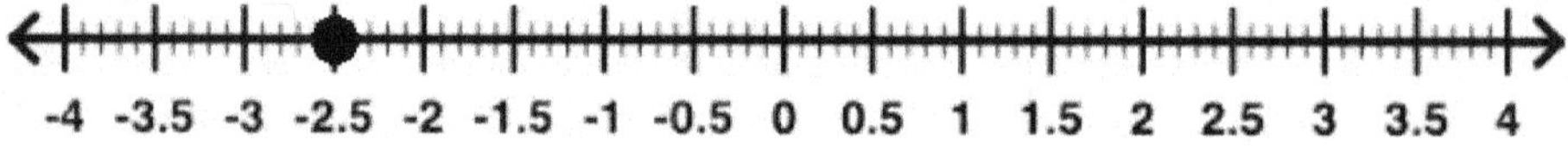

EXAMPLE: What is the opposite of -0.32?

Solution: The definition of an opposite number holds for the decimal numbers. Here the opposite of a negative number is a positive number. Thus, we get the opposite of - 0.32 is 0.32

Answer: **0.32**

2.6A - Opposite Numbers

351) What is the opposite of 2?

352) What is the opposite of -2?

353) What is the opposite of -122?

354) What is the opposite of 56?

355) What is the opposite of 810?

2.6A - Opposite Numbers

356) What is the opposite of -10.52?

357) What is the opposite of 0?

358) What is the opposite of -109?

359) What is the opposite of 8.17?

360) What is the opposite of -0.32?

2.6A - Opposite Numbers

361) Graph the opposite of -(-2.5) on the number line.

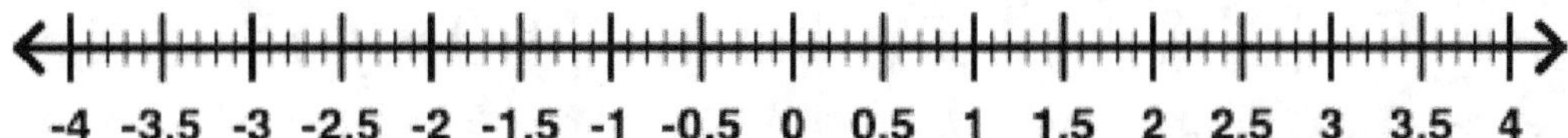

362) Graph the opposite of 5/2 on the number line.

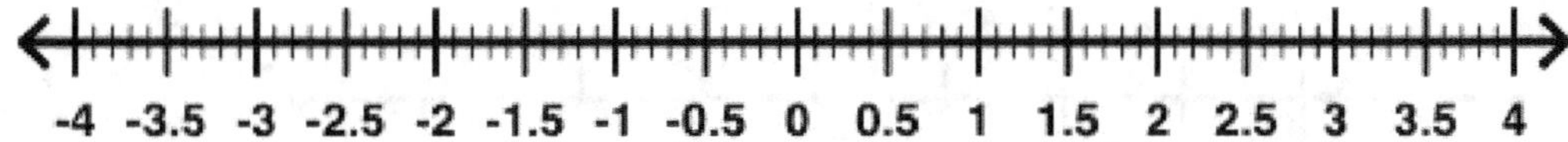

363) Which point represents the opposite of 0 on the number line?

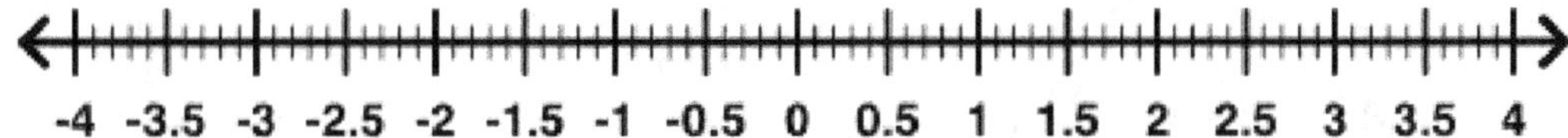

364) Which point represents the opposite of -5.5 on the number line?

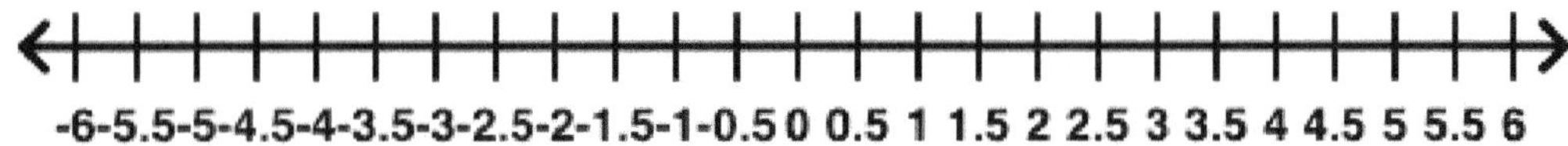

365) Graph the opposite $\frac{3}{2}$ on the number line.

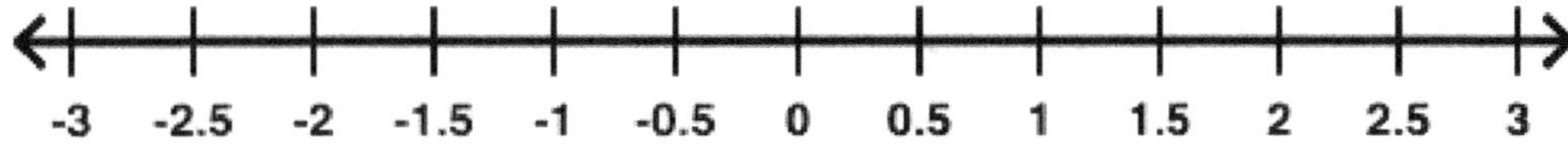

2.6A - Opposite Numbers

366) Which point represents the opposite of -(-1) on the number line?

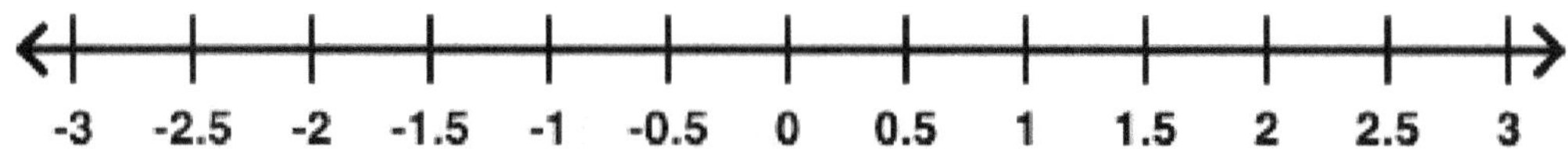

367) Locate the opposite of -1 on the number line.

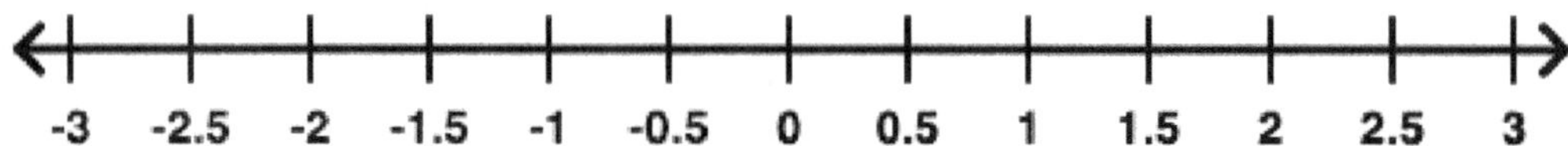

368) Locate the opposite of -2.2 on the number line.

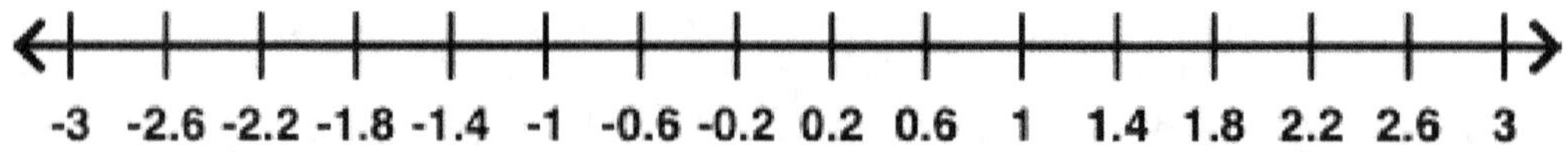

369) David made money by selling stocks last month. His investment income was $401. His income is -$302. What does this mean?

370) Suppose you are swimming along the surface of the sea. We use positive numbers to represent elevations above the surface of the sea and negative numbers to represent depths under the surface of the sea. You are sitting on the deck of a ship one level above Dylan, and Samantha dove in the water beneath Dylan. Who is at an elevation of -3 m?

2.6B - Quadrants

A *quadrant* refers to the four quarters of the coordinate plane (a two-dimensional plane defined by two perpendicular axes). The horizontal axis is called the *x-axis*, while the vertical axis is called the *y-axis*. **EXAMPLE:** Which quadrant is Point A located in?

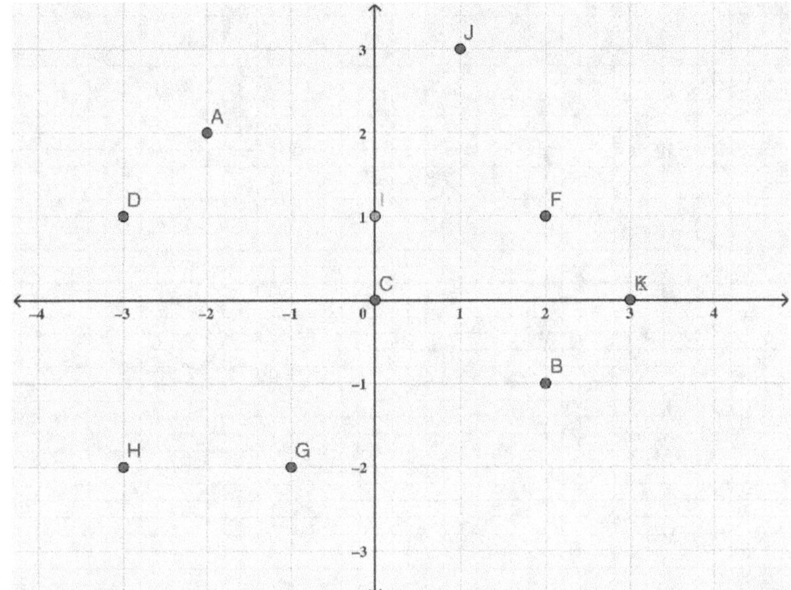

Solution: The quadrants start in the top right and count counterclockwise. Hence, Point F is in the 1st quadrant, Point A is in the 2nd quadrant, Point H is in the 3rd quadrant, and Point B is in the 4th quadrant.　　　　Answer: Point A is in the **2nd quadrant**

EXAMPLE: The Point H is given as (-3, -2). What point do you get when you reflect H across the x-axis then across the y-axis?

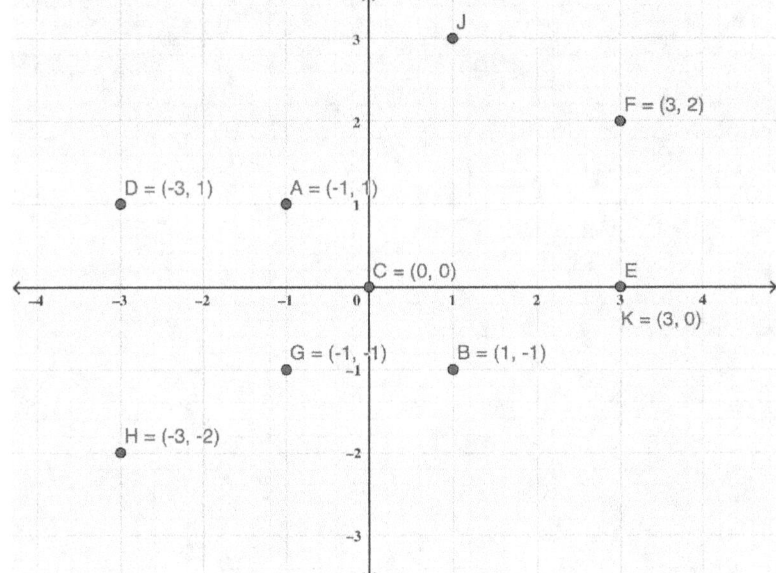

Solution: If H (-3, -2) is reflected across the x-axis, the obtained point is (3, -2). Then, if reflected across the y-axis, the obtained point is F (3, 2).　　　　Answer: **F (3, 2).**

2.6B - Quadrants

371) Which quadrant is Point A located in?

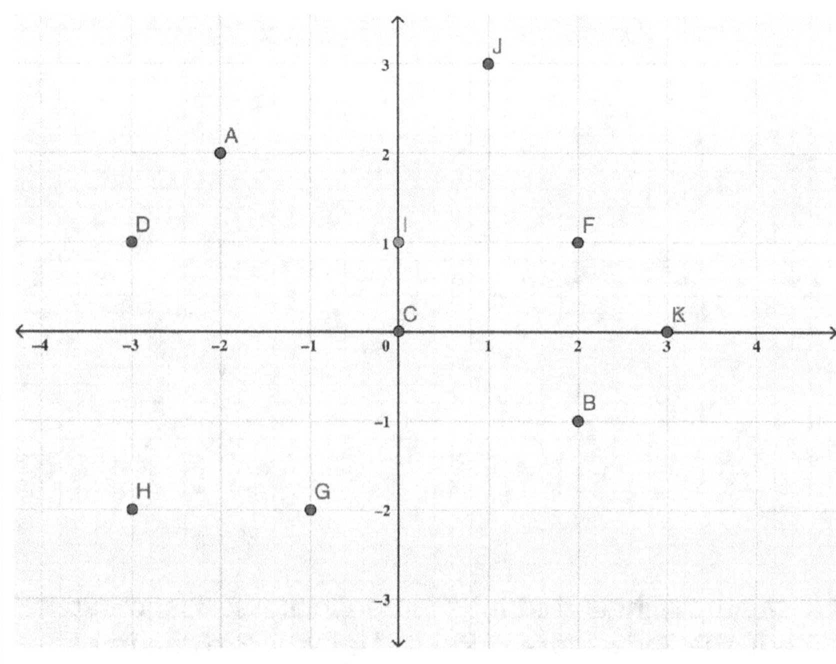

372) On the coordinate plane below, which point/s is/are in quadrant IV?

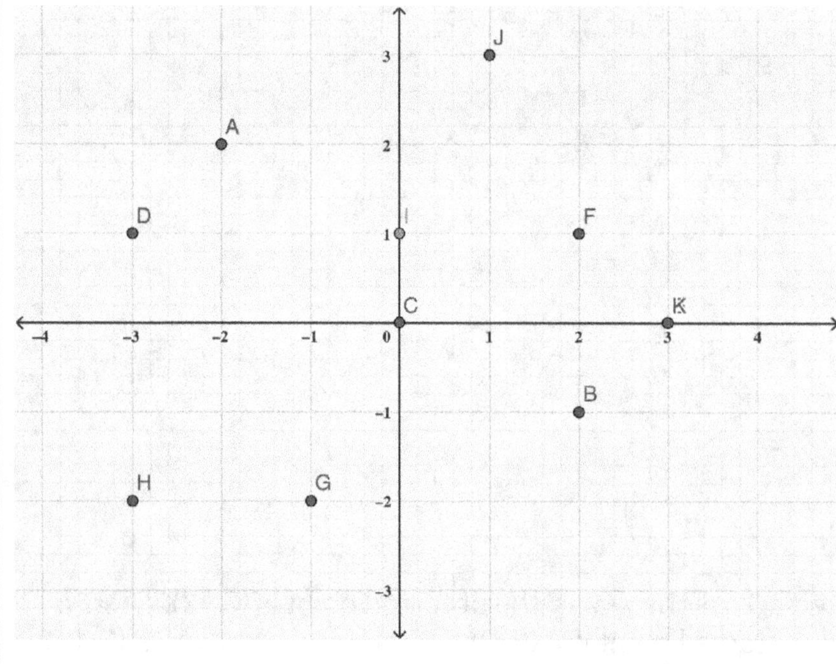

2.6B - Quadrants

---

373) Which quadrant is the ordered pair (-2, 2) located?

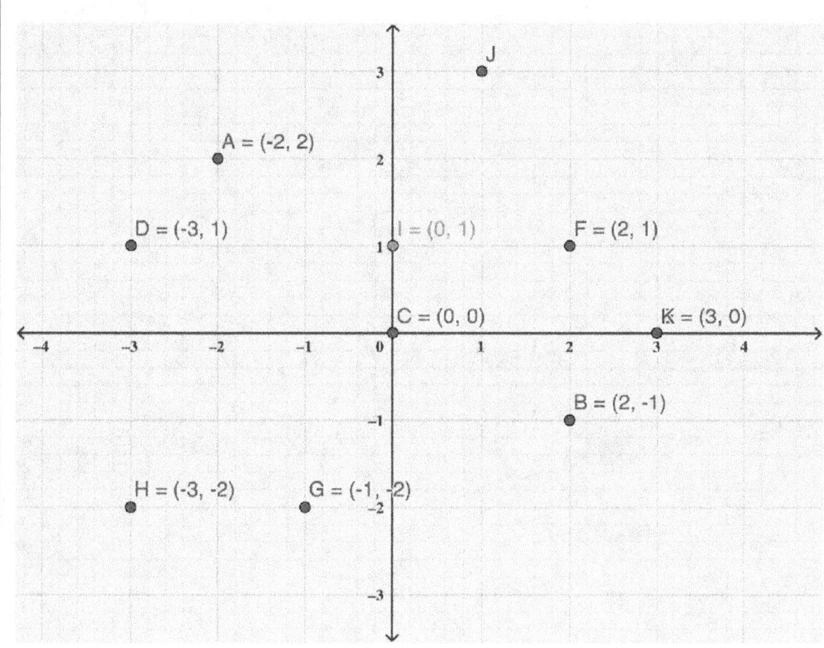

---

374) The Point H is given as (-3,-2). What point do you get when you reflect H across the x-axis then across the y-axis?

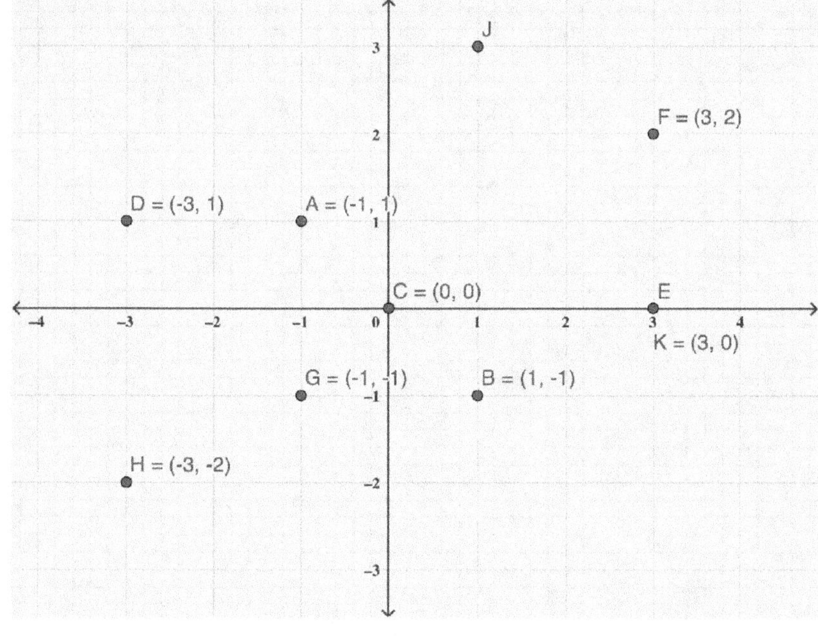

2.6B - Quadrants

---

375) The Point A is given as (-1,1). What point do you get when you reflect A across the y-axis then across the x-axis?

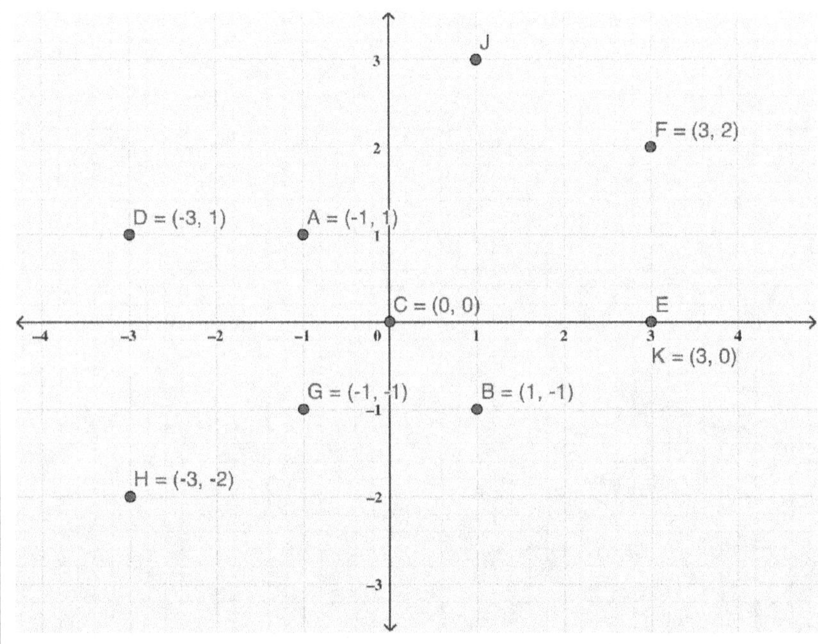

---

376) The two points (-3,-1) and (-3, -2) are on the coordinate plane below. Find the distance between these two points.

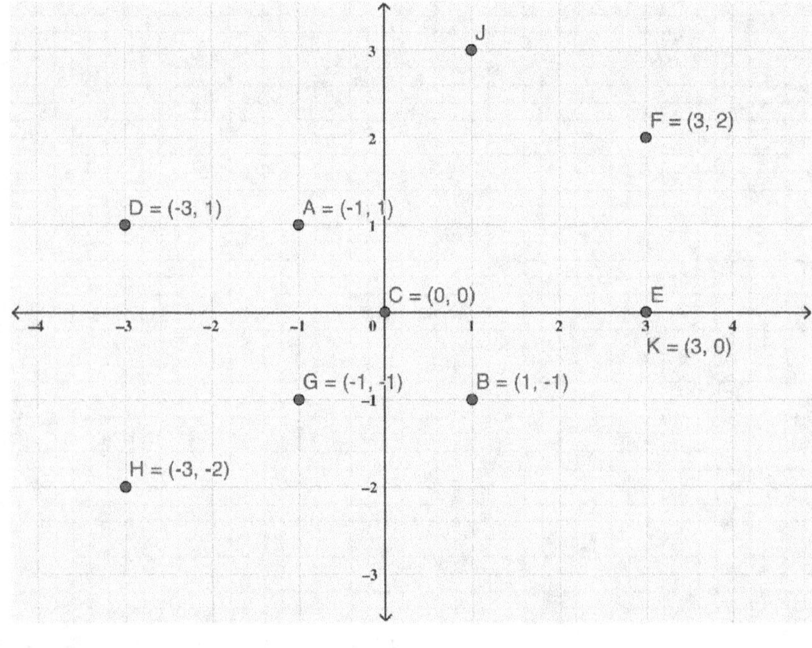

2.6B - Quadrants

377) Point D is located at (-3, 1). What point is three units below point D?

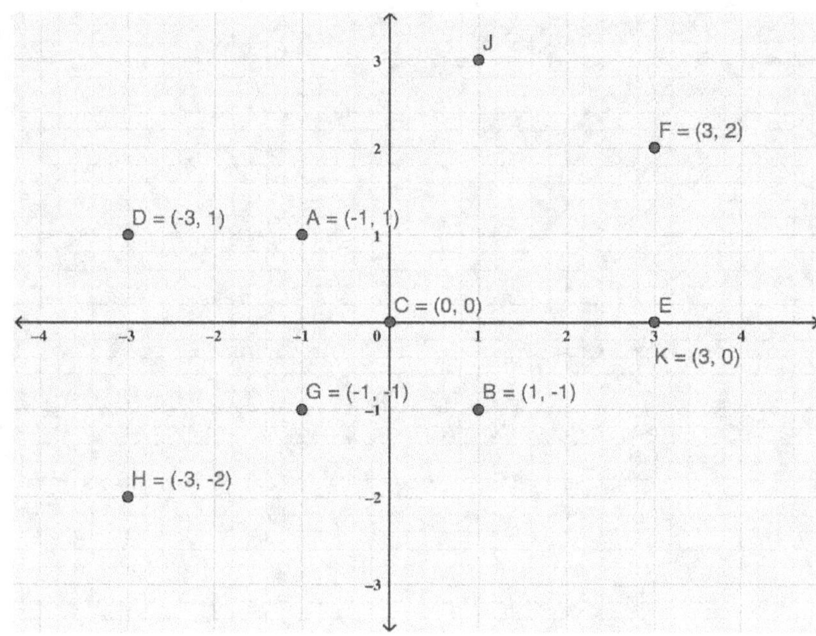

378) Point J is located at (1, 3). Which point is four units below Point J?

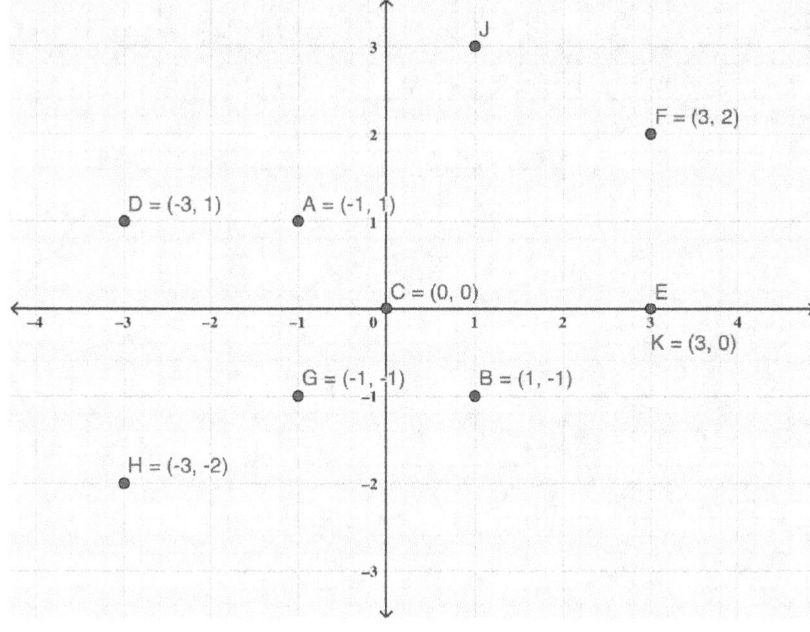

2.6B - Quadrants

---

379) Point G is located at (-1, -1). Which point is two units to the right of Point G?

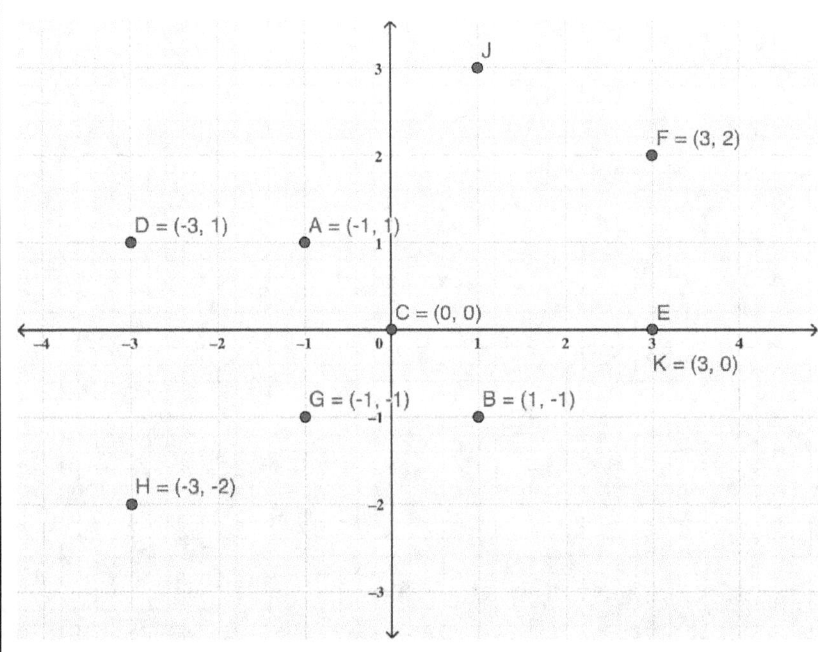

---

380) The two points (0, 0) and (3, 0) are on the coordinate plane below. Find the distance between these two points.

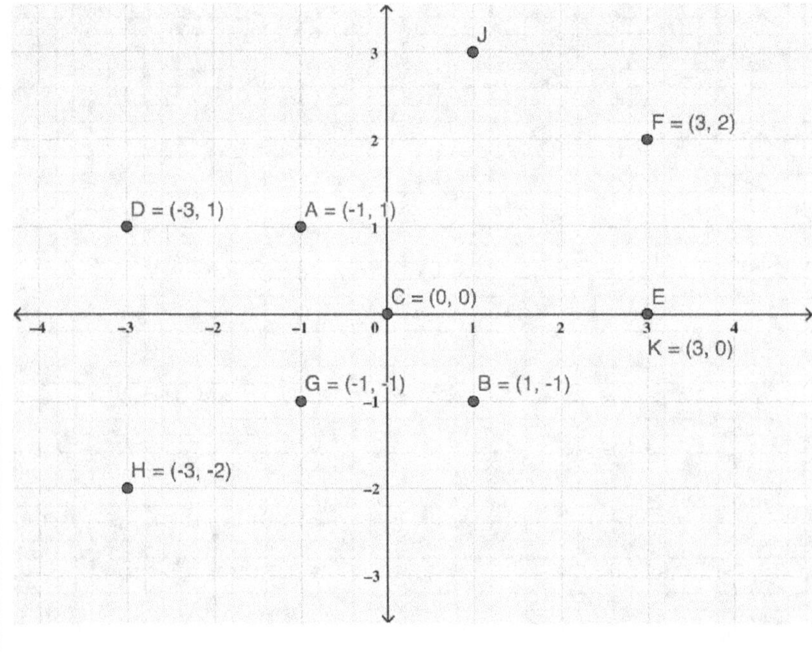

2.6C - Number Lines and Coordinates

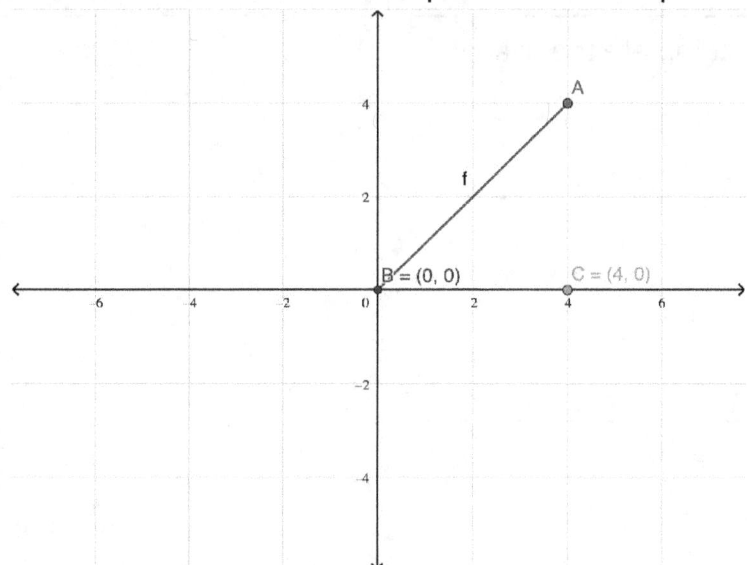

EXAMPLE: Which coordinate point would complete the triangle?

Solution: To complete the triangle, we need to draw a line connecting Point A and Point C. Thus, Point C (4, 0) would complete the triangle.

Answer: **Point C (4, 0)**

EXAMPLE: What is the area of the rectangle with these coordinates?
F (1, 4), G (1, 2), (4, 2) and (4, 4)

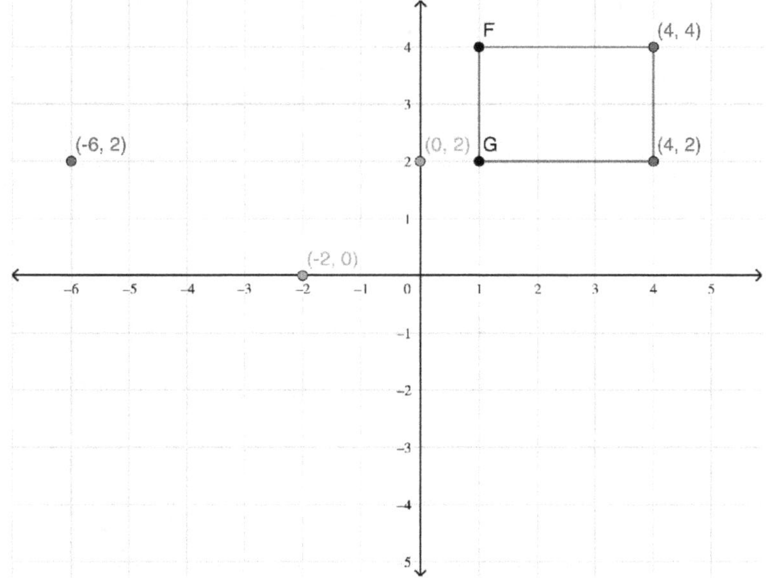

Solution: The length of the two sides of the rectangle are two units and three units.
If area = length × width, then area = 2 units × 3 units = 6 square units.

Answer: **6 square units**

2.6C - Number Lines and Coordinates

381) Which coordinate point would complete the triangle?

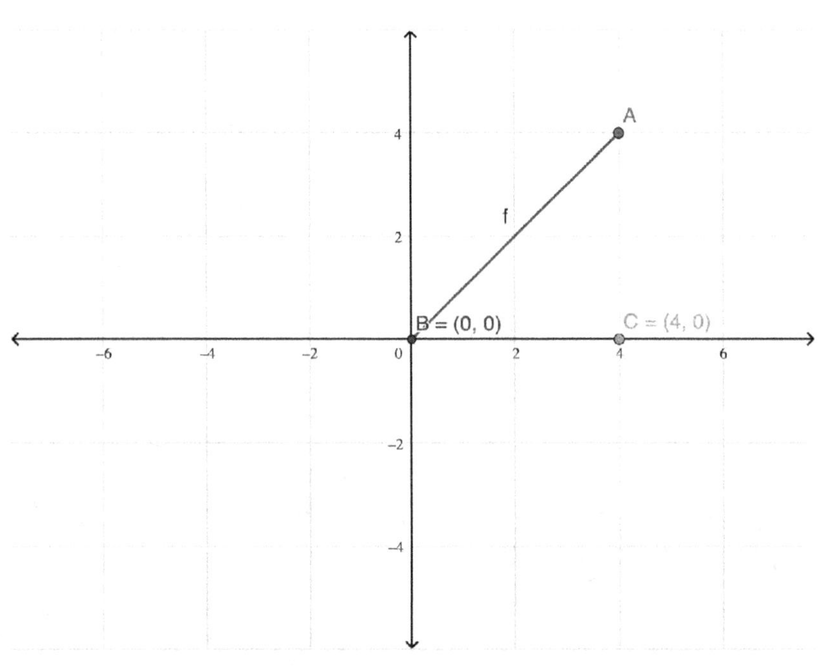

382) Name coordinate points of the shape.

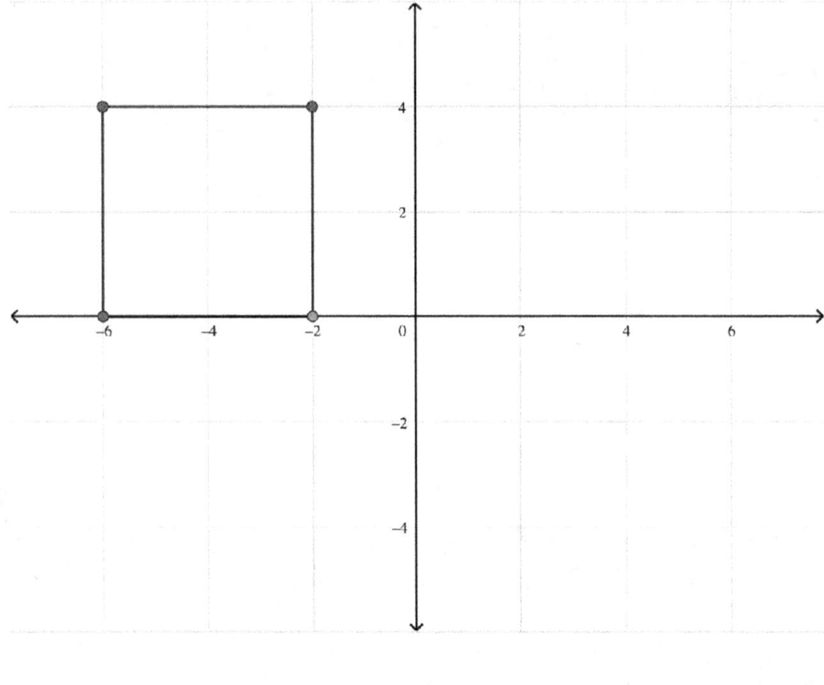

2.6C - Number Lines and Coordinates

383) Which coordinate point will complete the square?

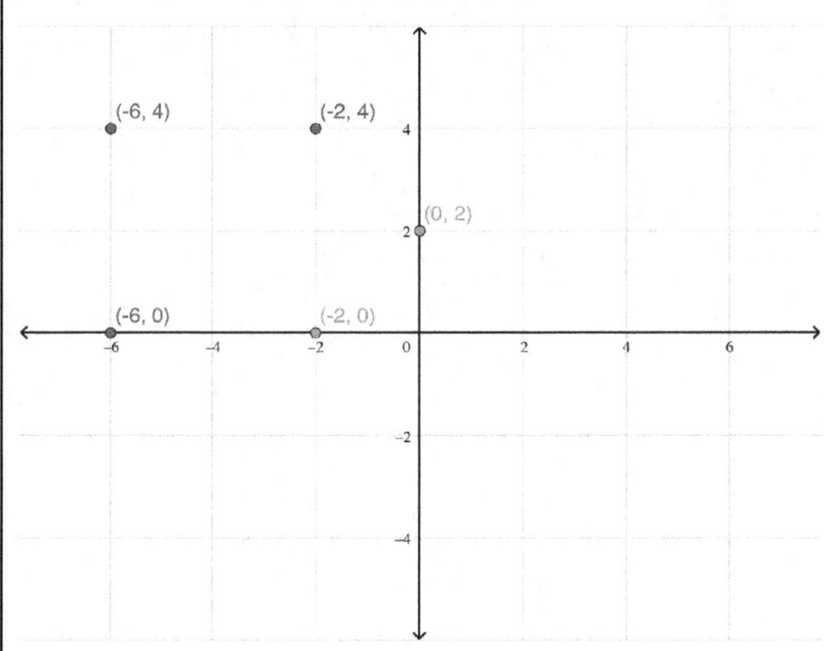

384) Which coordinate point will complete the rhombus?

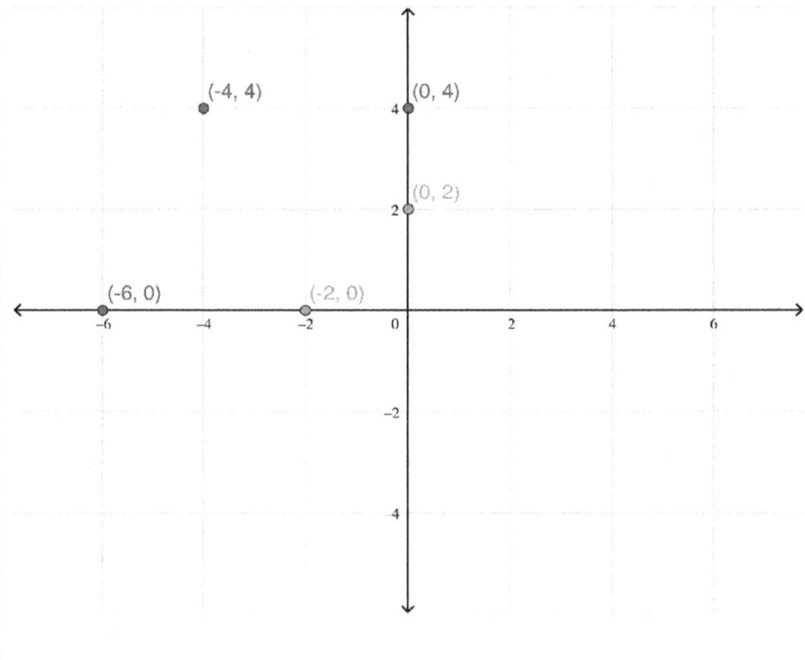

2.6C - Number Lines and Coordinates

385) What is the area of the triangle with these coordinates?
(-4, 4), (-6, 0) and (-2, 0)

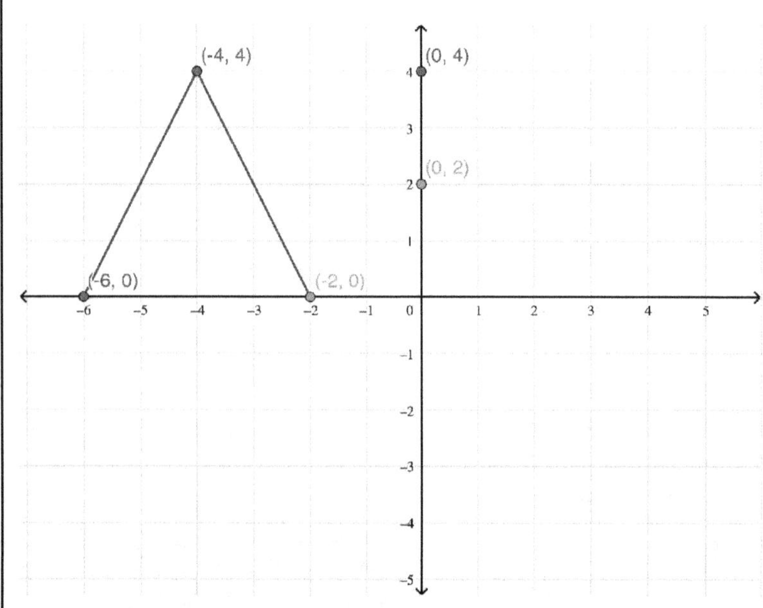

386) What is the area of the rectangle with these coordinates?
F (1, 4), G (1, 2), (4, 2) and (4, 4)

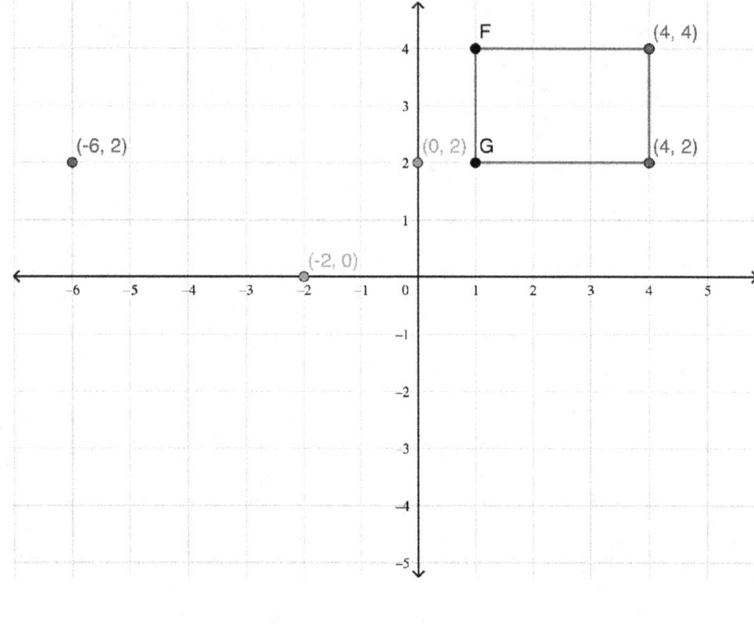

2.6C - Number Lines and Coordinates

387) Name the coordinates of this polygon.

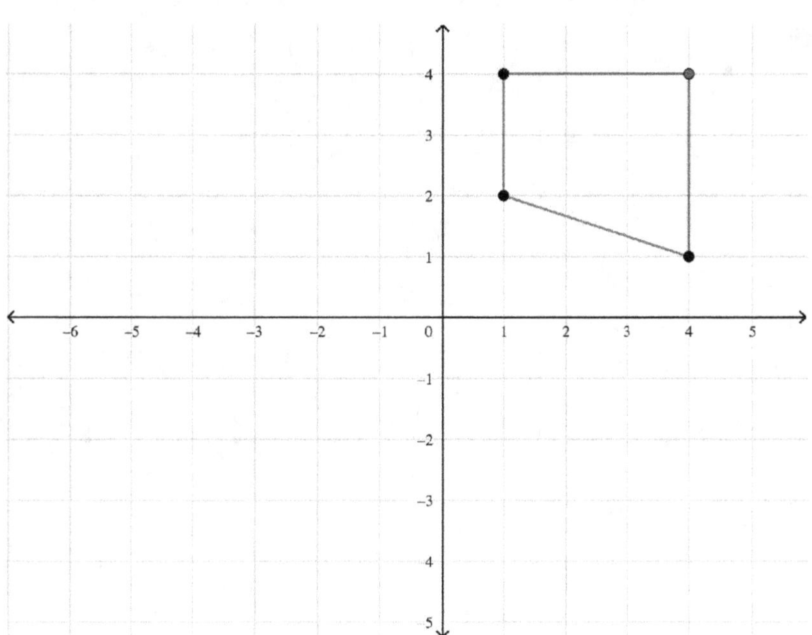

388) What is the area of the polygon with these coordinates?

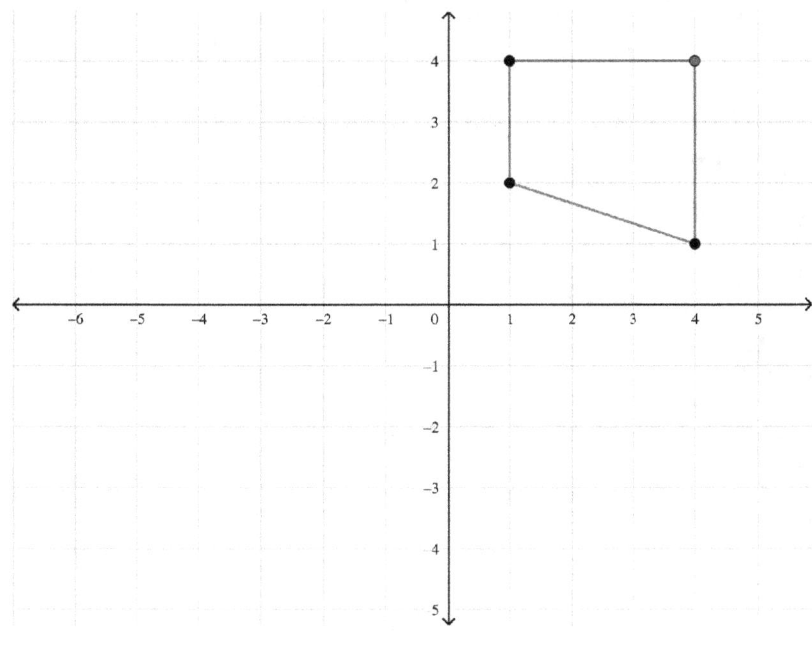

2.6C - Number Lines and Coordinates

389) What is the area of the polygon with these coordinates?

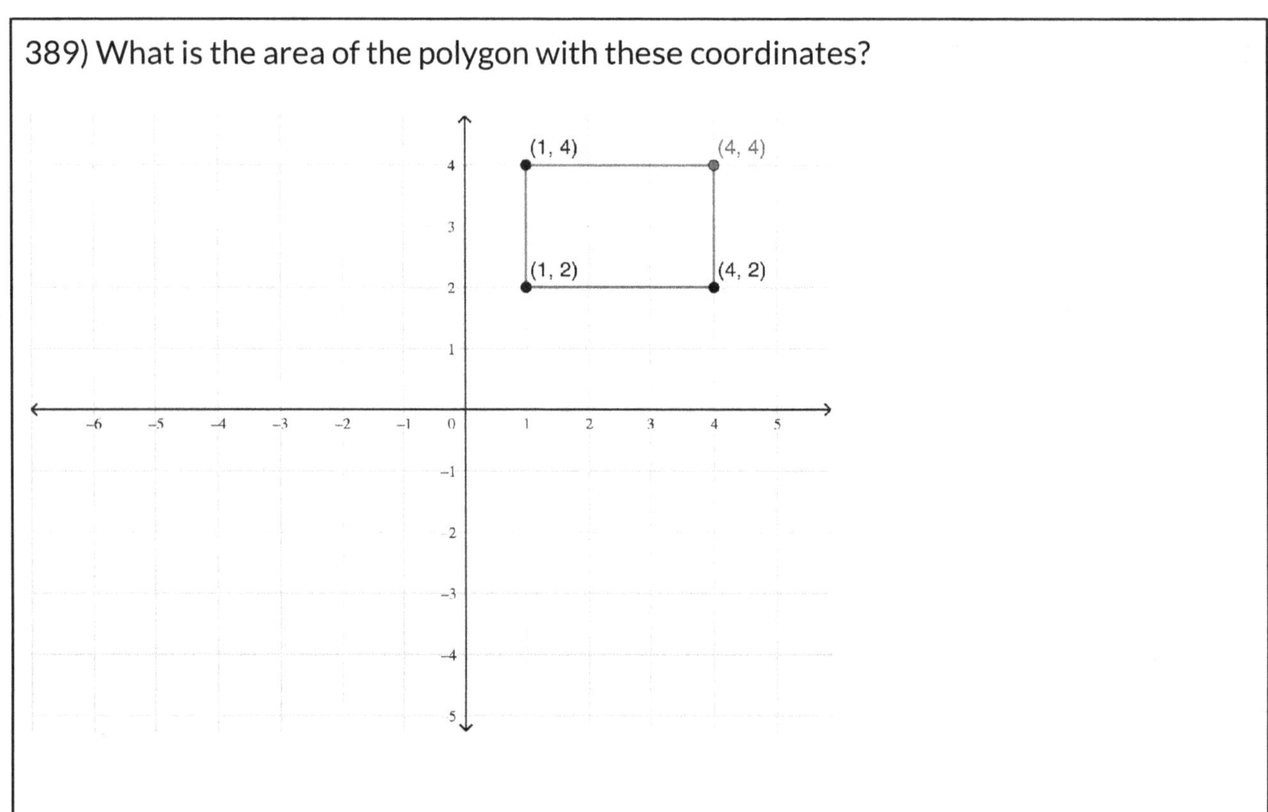

390) How many students like pizza?

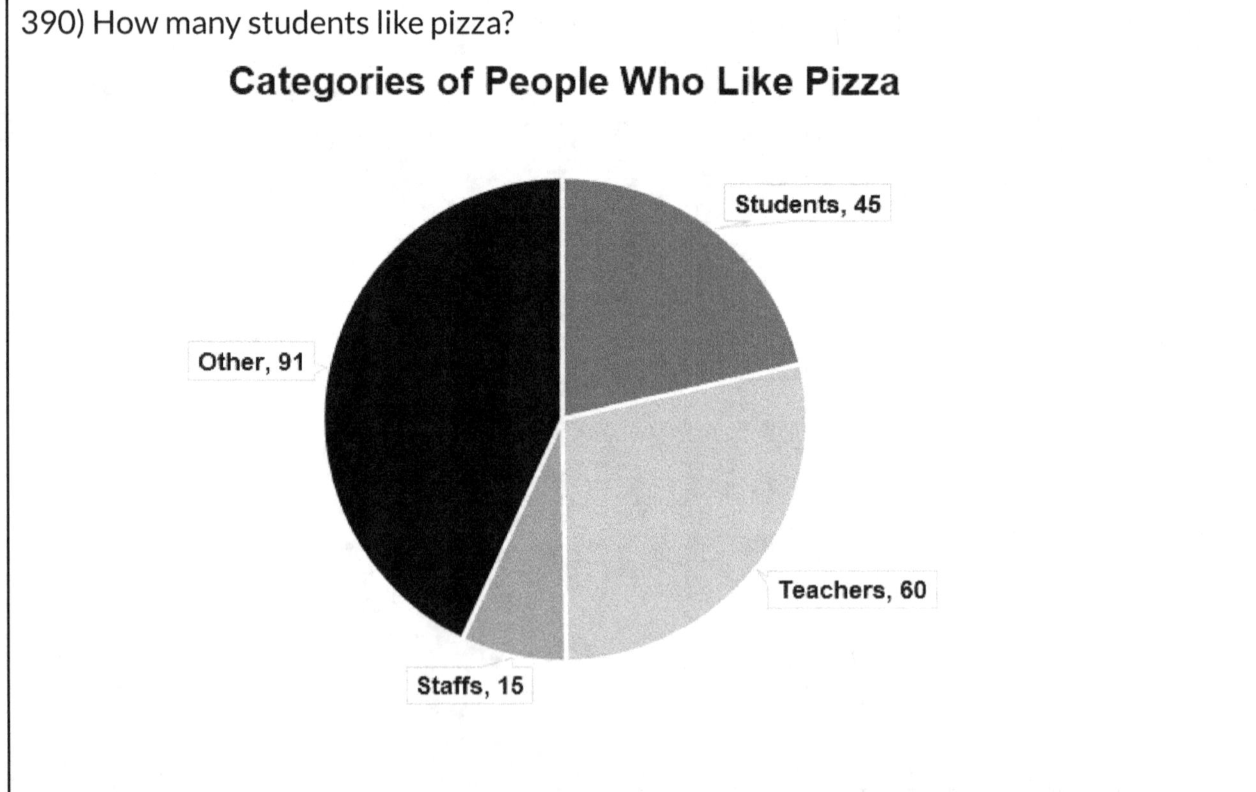

2.6C - Number Lines and Coordinates

391) How many have ages not included in the range of 12 to 14??

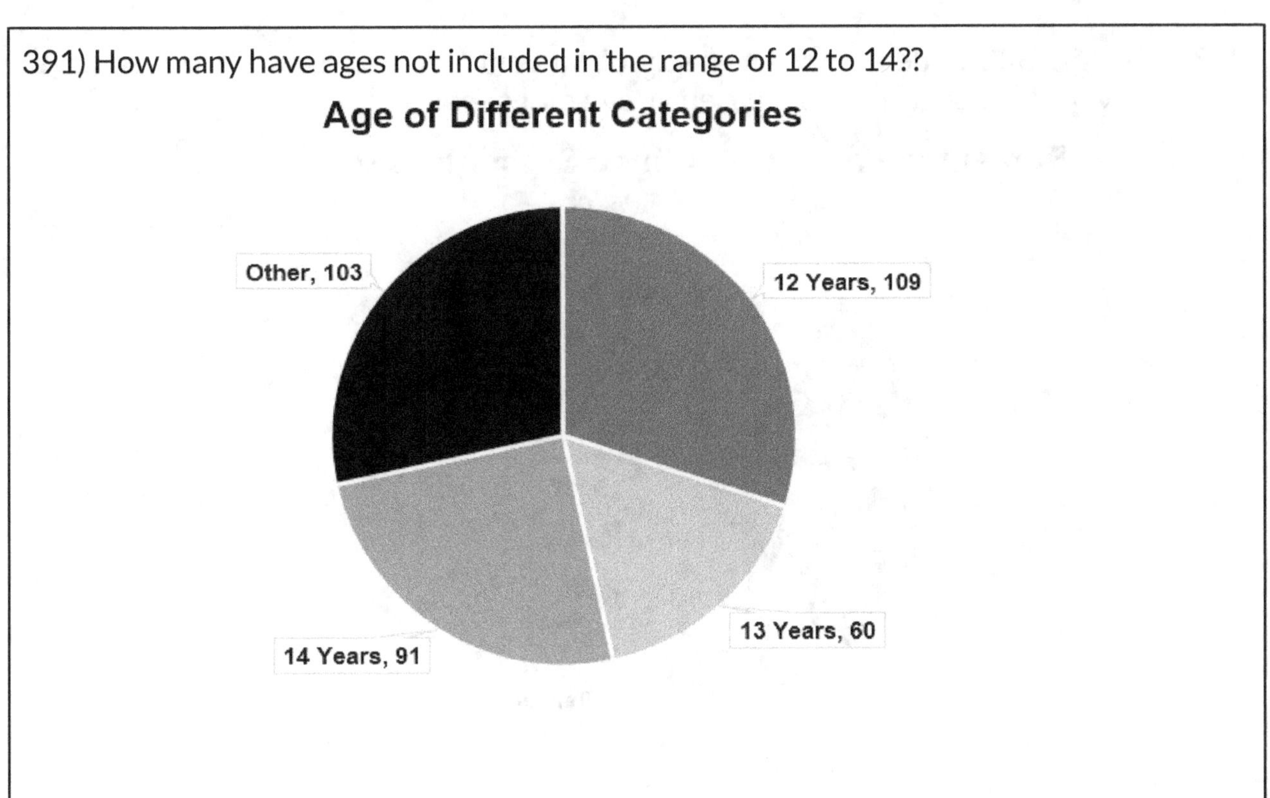

392) Which month was the coldest month of the year 2011?

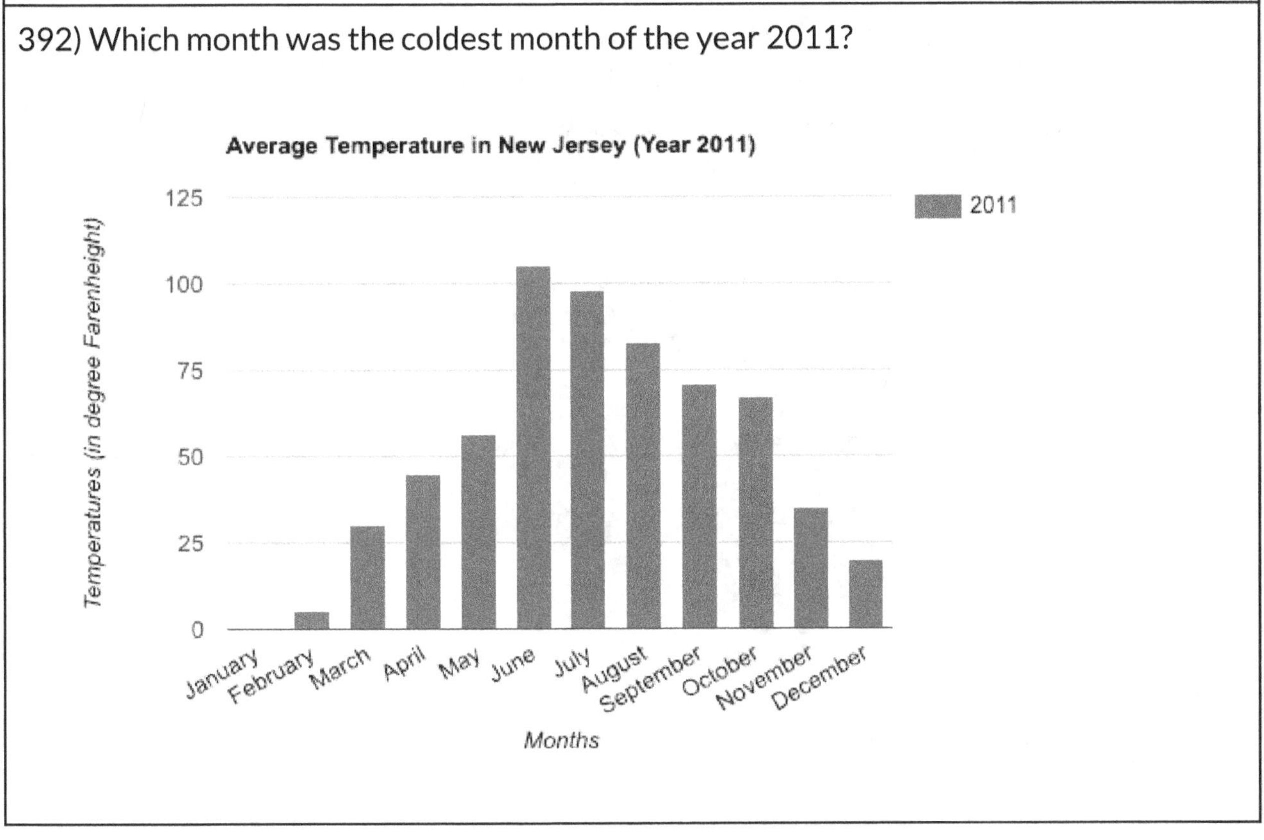

2.6C - Number Lines and Coordinates

393) There are 500 students in the high school. Students are represented in percentage as they like football, soccer, tennis, baseball, and other sports.

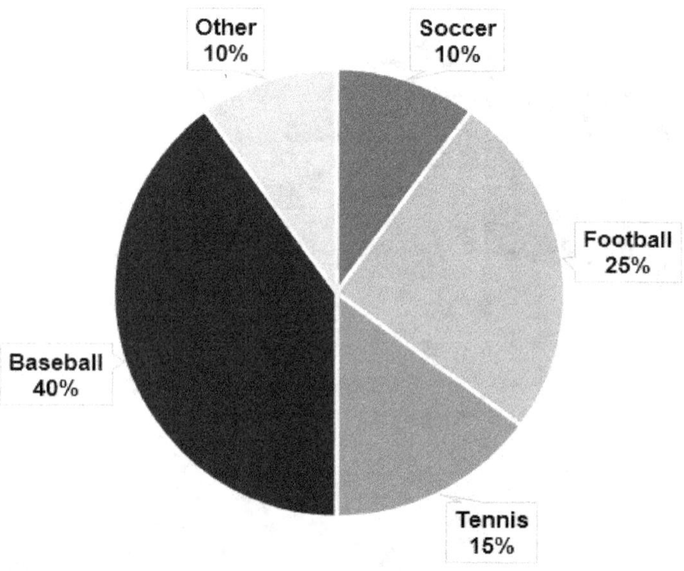

How many of them like tennis?

394) There are 500 students in the high school. Students are represented in percentage as they like football, soccer, tennis, baseball, and other sports.

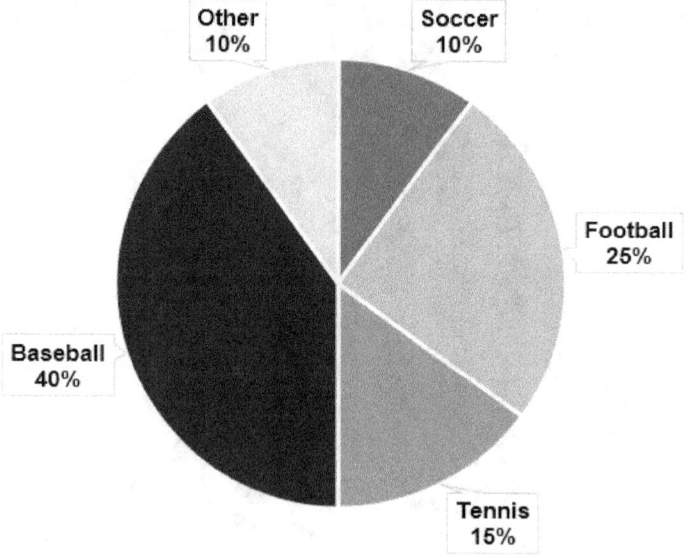

How many of them like soccer?

2.6C - Number Lines and Coordinates

395) There are 500 students in the high school. Students are represented in percentage as they like football, soccer, tennis, baseball, and other sports.

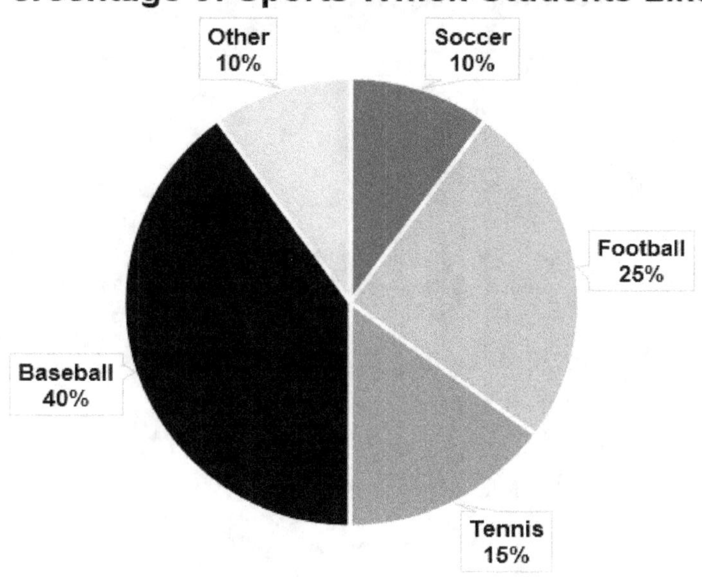

How many of them like baseball?

396) Which month was the warmest month of the year 2011?

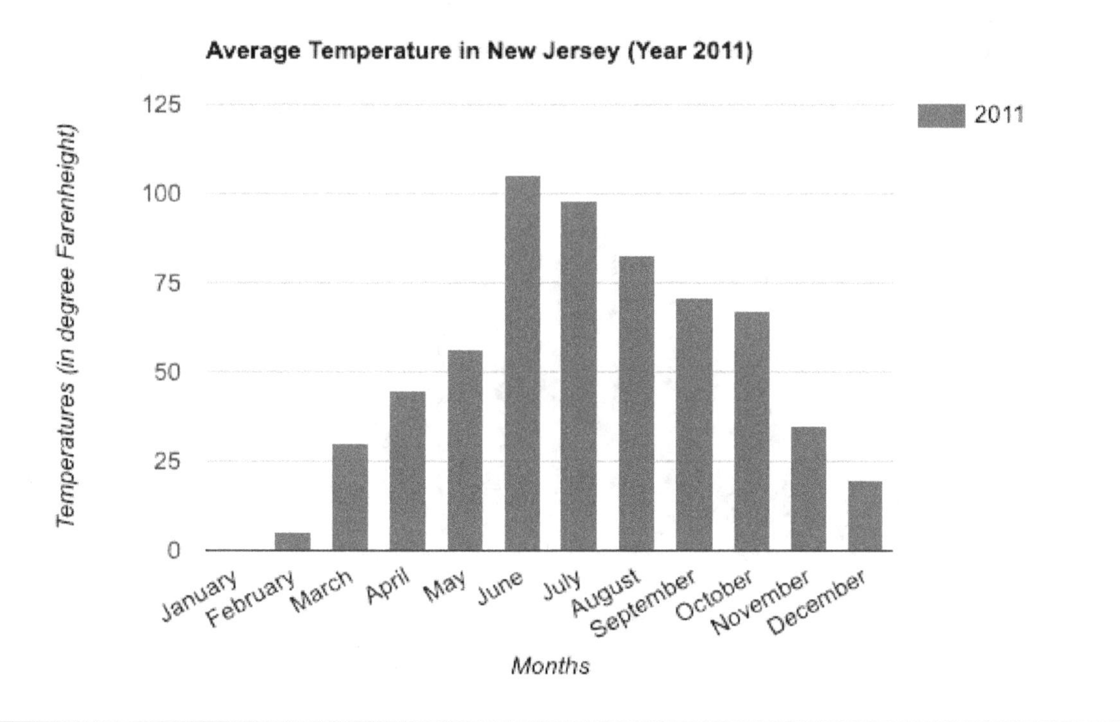

2.6C - Number Lines and Coordinates

397) Arrange the months from warmest to coldest.

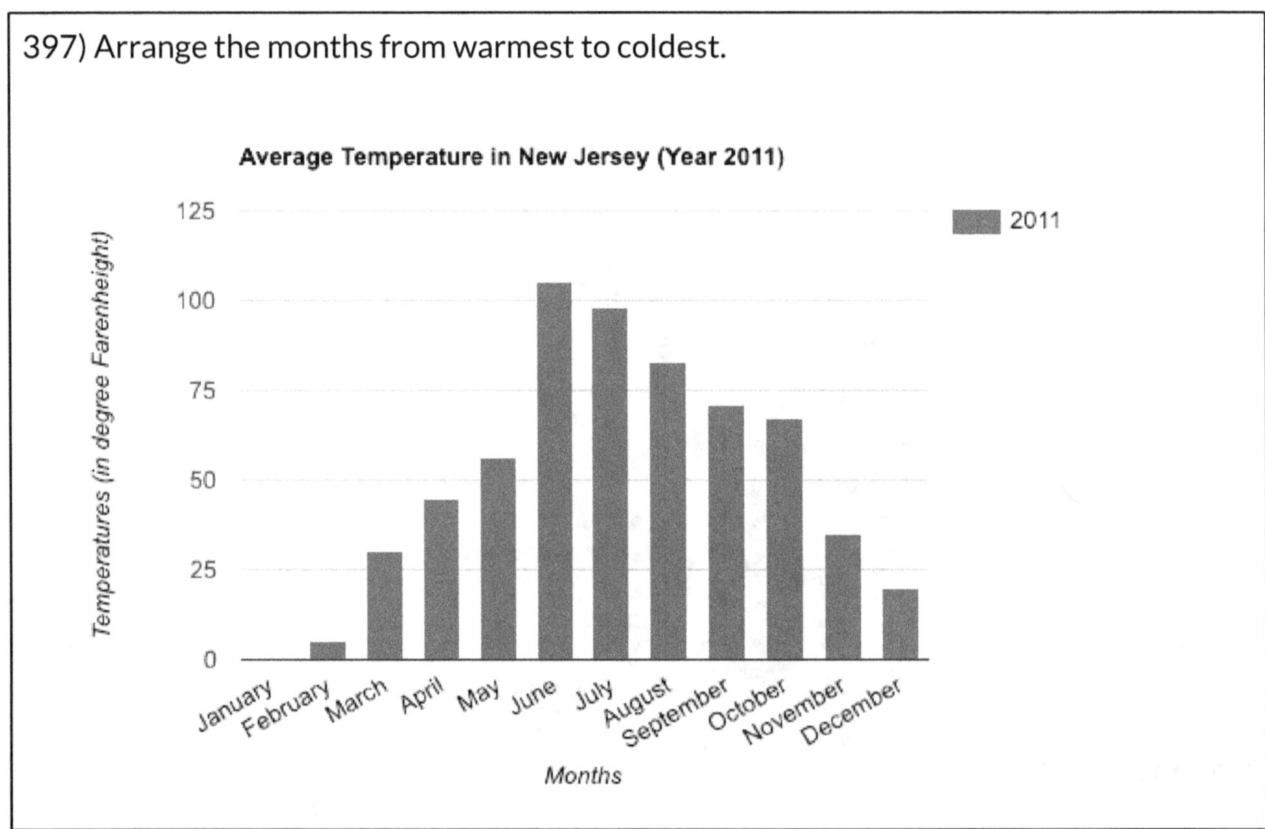

398) What is the fourth hottest month in 2011?

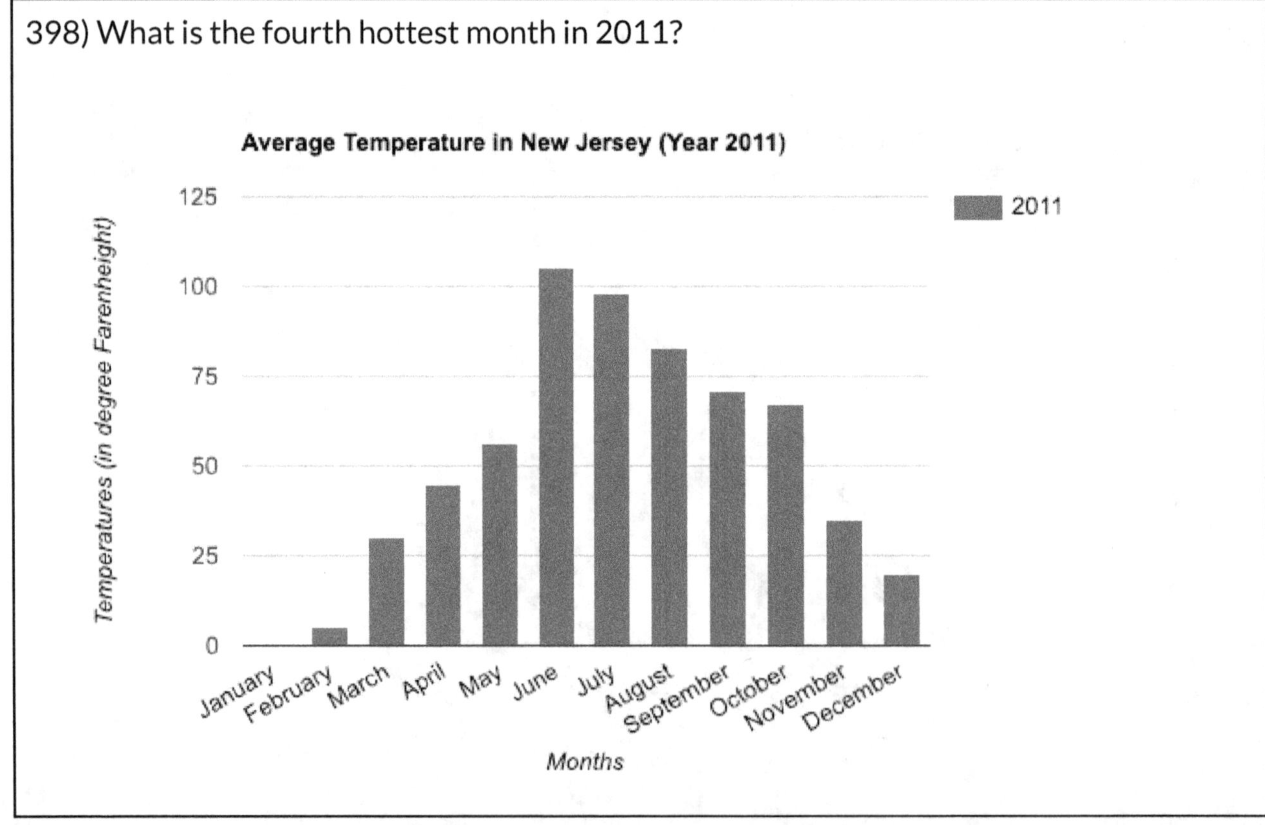

2.6C - Number Lines and Coordinates

---

399) Which points of the y-coordinate on the grid below are smaller than 4?

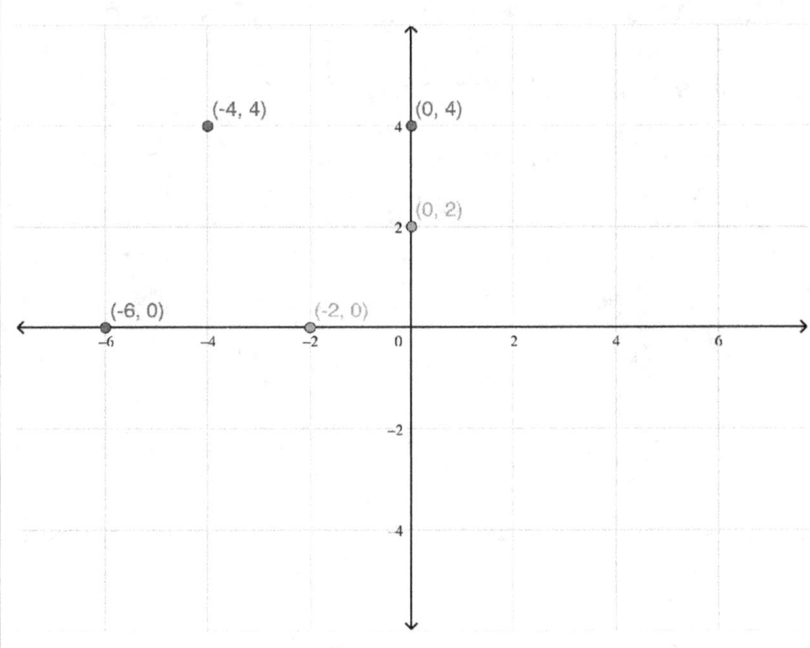

---

400) What is the ordered pair for the following grid?

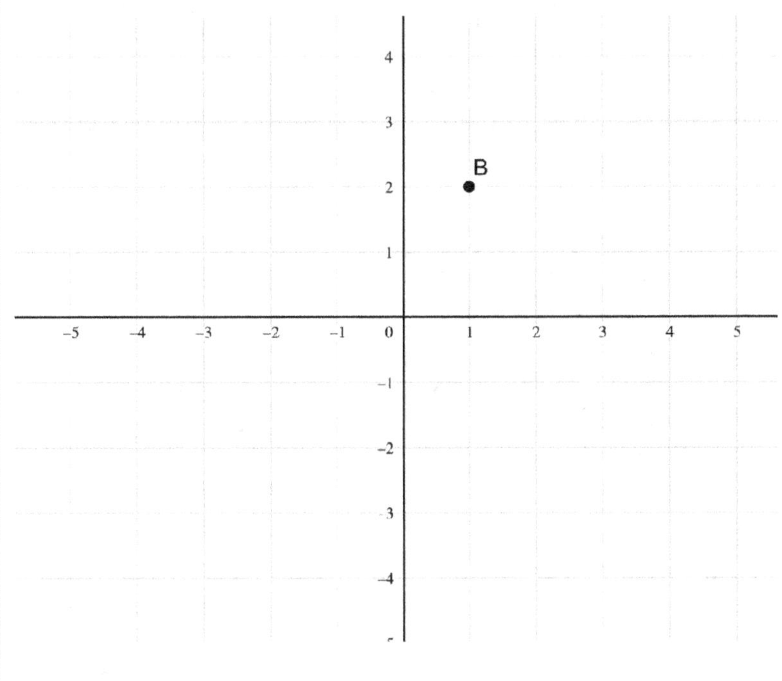

2.7A - Compare Integers and Number Order

In comparing integers, negative numbers are always lesser than zero and positive numbers (e.g., -3 is less than 0 and 3). In comparing two positive integers, the larger the number, the greater its value (e.g., 8 is greater than 4). In comparing two negative numbers, the larger its absolute value, the lesser its actual value (e.g., -10 is less than -5).

EXAMPLE: Order the following numbers from least to greatest: -2, 5, -3, 0.

Solution: We look for the numbers that are lowest to the highest. From -2, 5, -3, and 0, the lowest number is -3, and the highest number is 5. Thus, we write -3, -2, 0, 5.

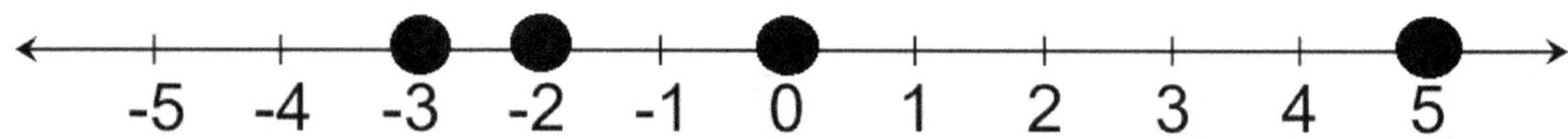

Answer: **-3, -2, 0, 5**

EXAMPLE: How do you interpret -10 < 7?

Solution: To compare two numbers, we use *less than* (<), *greater than* (>), or *equal* (=). Hence, we interpret -10 < 7 as the number -10 is less than the number 7.

Answer: **-10 is less than 7**

EXAMPLE: Write an inequality that correctly compares -10 and -32.

Solution: We conclude that -32 is less than -10. Thus, we write **-32 < -10**. We can also conclude that -10 is greater than - 32. Hence, we can also write **-10 > -32**.

Answer: **-32 < -10** or **-10 > -32**

2.7A - Compare Integers and Number Order

---

401) Is 3 greater than -10 for this number line?

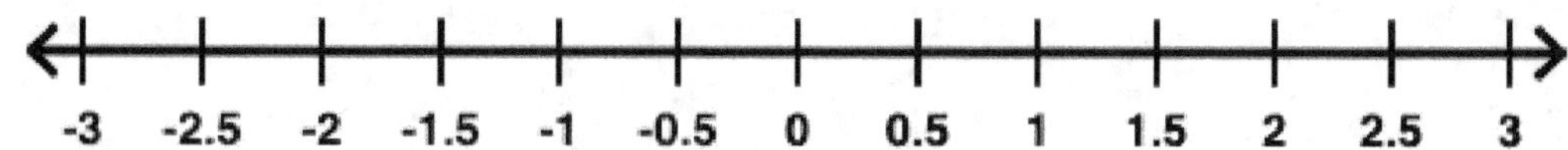

---

402) Is 0 greater than -3 for this number line?

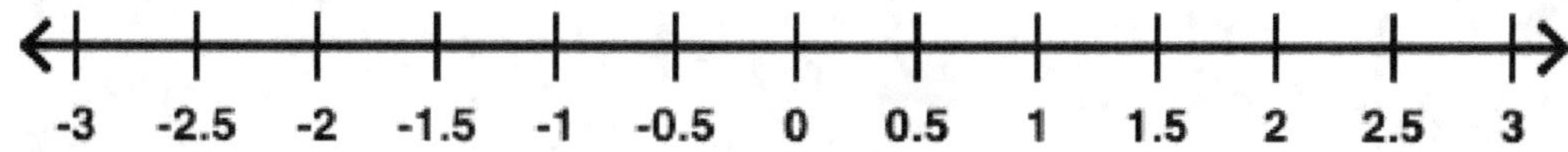

---

403) Order the following numbers from least to greatest: -2, 5, -3, 0.

---

404) Order the following numbers from least to greatest: -7, -1, -3, -9.

---

405) Order the following numbers from least to greatest: -12, 15, -13, 0, 19.

2.7A - Compare Integers and Number Order

406) Order the following numbers from least to greatest: -2, 15, 0, 7, 10.

407) Order the following numbers from least to greatest: 0, 22, -12, 15, -13, 18.

408) Rearrange the following numbers from least to greatest: 0, -10, 14, -90.

409) Rearrange the following numbers from least to greatest: 12, -4, -50, 71, -109.

410) Rearrange the following numbers from least to greatest: 95, -10, 12, -49.

2.7A - Compare Integers and Number Order

411) Rearrange the following numbers from least to greatest: -18, 34, -87, -15.

412) How do you interpret 5 > - 4?

413) How do you interpret -10 < 7?

414) What is the relation between -10 and 4?

415) Select the true statements about the number line.

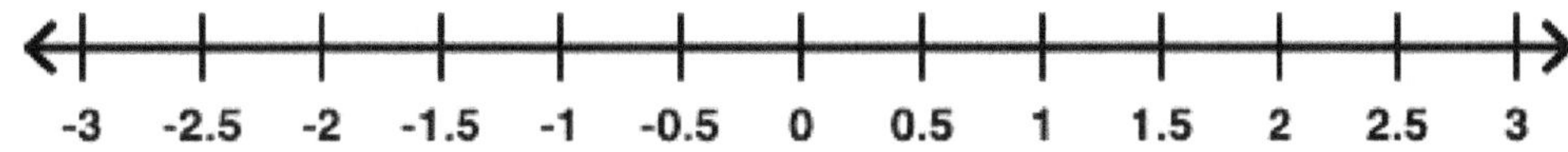

A. -3 > 0   B. 0 > -2

2.7A - Compare Integers and Number Order

416) Select the true statement about the number line.

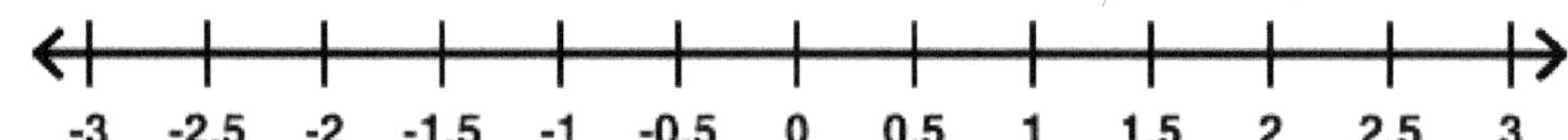

B. -3 > 3   C. 3 > -2

417) Write a relation that correctly compares 3 and 3.

418) Write an inequality that correctly compares 0 and -32.

419) Write an inequality that correctly compares 5 and 7.

420) Write an inequality that correctly compares -10 and -32.

2.7B - Compare and Order Word Problems

EXAMPLE: The temperature in New York is -12°C, and the temperature in Montana is -5°C. Which city is warmer?

Solution: Since the number -5 is greater than the number -12, consider the higher temperature between the two places is the warmer one. Now, the temperature in Montana is higher than the temperature in New York. Thus, the city of Montana is warmer.

Answer: **Montana**

EXAMPLE: Dylan cuts a cake into five different sizes:

$\frac{1}{10}, \frac{2}{5}, \frac{1}{20}, \frac{3}{10}, \frac{3}{20}$. What is the largest size of the cake?

Solution: Convert the fractions into similar fractions to identify their values easily: $\frac{2}{20}$, $\frac{8}{20}, \frac{1}{20}, \frac{6}{20}, \frac{3}{20}$ Here, the lowest value of the fraction is $\frac{1}{20}$ and the highest value of the fraction is $\frac{8}{20}$.

Hence, the largest size of the cake is $\frac{2}{5}$.

Answer: $\frac{2}{5}$

EXAMPLE: Bill has $50 in his hand. Now he gives $25 to Simon, $7 to Kelly, and $18 to Rachel. Who receives the most money?

Solution: The largest amount amongst $25, $7, and $18 is $25. Thus, Simon receives the most money.

Answer: **Simon**

2.7B - Compare and Order Word Problems

421) Order the following rational numbers: $-\frac{1}{7}, \frac{2}{5}, -\frac{1}{20}, 1\frac{3}{11}, 4\frac{4}{7}$	422) Order the following rational numbers: $\frac{3}{7}, \frac{2}{5}, \frac{1}{2}, -\frac{3}{10}, -\frac{7}{8}$
423) Four temperatures are given as 30°F, 50°F, 90°F, and 20°F. Arrange these temperatures from warmest to coolest.	424) The temperature in New York is -12°C, and the temperature in Montana is -5°C. Write down the inequality that correctly describes the temperature (warmer to cooler).
425) The temperature in New City was -2°C. In Macomb, the temperature was 3°C. In which city was the weather warmer? In which city was the weather cooler?	426) Temperatures of Brooklyn, Chicago, Dallas, Austin are -9°C, 0°C, 23°C, and 11°C, respectively. In which city is the weather coldest? In which city is the weather warmest?
427) Sandi recorded a temperature of 70°F. Bella recorded a temperature of 40°F. Who recorded the warmer temperature?	428) The temperature in New York is -12°C, and the temperature in Montana is -5°C. Which city is colder?
429) Dillion has $50 in his hand. Now he gives $25 to Simon, $7 to Tulip, and $18 to Rachel. Who receives the least money?	430) Susan cuts a cake into five different sizes: $\frac{1}{10}, \frac{2}{5}, \frac{1}{20}, \frac{3}{10}, \frac{3}{20}$. What is the smallest size of the cake?

2.7B - Compare and Order Word Problems

431) Gabriel and Gomez are making pancakes. Gabriel uses 3 lbs of flour to form his pancake, and Gomez uses 9.5 lbs. of flour. Who is making the larger pancake?	432) Trina runs $\frac{3}{10}$ miles on Monday, $\frac{2}{5}$ mile on Tuesday, $\frac{5}{9}$ miles on Sunday, and $\frac{2}{3}$ miles on Friday. On what day does she run the most distance?
433) David walks $\frac{1}{3}$ miles on Saturday, $\frac{5}{6}$ mile on Sunday, $\frac{7}{8}$ miles on Monday, and $\frac{9}{5}$ miles on Friday. On what day does he walk the most distance?	434) The area of a triangle is $4\frac{1}{10}$ square meters. The area of a rectangle is $2\frac{1}{10}$ square meters. Which shape has a greater area?
435) The electric charge of sodium is +1. The electric charge of calcium is +2. Which one has a more positive electric charge?	436) Maggy spends $25 for a bouquet of flowers, James spends $32, while Simon spends $20. Who spends the least money?
437) The volume of a cone is nine cubic meters. The volume of a rectangular prism is 13.5 cubic meters. Which shape has a greater volume?	438) An airplane is flying at an elevation 20 meters above the ground. A bird is flying at an elevation 18 meters above the ground. Write an inequality that correctly compares the elevations.
439) The area of a rhombus is 8/9 square centimeters. The area of a rectangle is $2\frac{1}{5}$ square centimeters. Which shape has a greater area?	440) The volume of a sphere is 15.6 cubic meters. The volume of a cube is 25.2 cubic meters. Which shape has a greater volume?

2.7C - Absolute Value

An *absolute value* refers to the distance of a number on the number line from 0 without considering its direction from zero where a number lies. As distance is always positive, an absolute value of a number is never negative.

EXAMPLE: What is the absolute value of -10?

Solution:
| -10 | = 10. The absolute value of -10 is 10. This means that -10 is ten units away from 0 on the number line.

Answer: **10**

EXAMPLE: Order the following values from least to greatest: |80|, -23, |-56|, 23.

Solution: The absolute value of 80 is 80. The absolute value of -56 is 56. Thus, write the values from least to greatest as -23, 23, 56, 80.

Answer: **-23, 23, |-56|, |80|.**

EXAMPLE: Jake got penalized a -8 point. Joshua got penalized a -10 point. Who got penalized the lesser?

Solution: The absolute value of -8 is 8. The absolute value of -10 is 10. Since 10 is greater than 8, Jake got penalized the lesser.

Answer: **Jake**

2.7C - Absolute Value

441) What is the absolute value of -1?	442) What is the absolute value of -15?
443) What is the absolute value of -7.15?	444) What is the absolute value of $-\frac{1}{5}$?
445) What is $|-209|$?	446) What is $\left\|-\frac{7}{9}\right\|$?
447) What is $\left\|8\frac{7}{9}\right\|$?	448) What is $|-10|$?
449) Order the following values from least to greatest: $|80|$, -23, $|-56|$, 23.	450) Order the following values from least to greatest: 4, -42, $|-16|$, $|53|$, 9.

2.7C - Absolute Value

451) If two numbers are -0.12 and 1.12. Which number has the greatest absolute value?	452) If two numbers are -190 and 410. Which number has the greatest absolute value?
453) If two numbers are 40 and -12. Which number has the greatest absolute value?	454) If two numbers are -90 and 40. Which number has the greatest absolute value?
455) What is the greatest absolute value of -9.1 and 12.3?	456) Joshua got penalized a -7. Jake got penalized a-10. Who got penalized the lesser?
457) Sally's account balance is $32.89. Noah's account balance is -$5. Who has more debts?	458) The temperature of the water is 100°F. The temperature of the oil is 98°F. Which is warmer?
459) Joshua got a 94% on his math test, while Lucas got a 92% on his. Who got the higher score?	460) A dolphin swims at -20.5 meters. A bird flies at 11.9 meters from the ground. Which animal is closer to sea level?

2.7C - Absolute Value

461) An account balance is -$45.6. What is the absolute value of -$45.6?	462) Joshua scored 79.1. Jake scored 81.9. Who scored the highest?
463) A pen has a value of $7.9, a pencil has a value of $4.5, and an eraser has a value of $1. Samantha bought the pencil and the eraser. Sally bought a pen. Who spent the least money?	464) Compare $|-10|$ and $|-2|$.
465) Compare $|-6|$ and $|4|$.	466) $|-10| + 5 = ?$
467) $|-23| + 15 = ?$	468) The temperature of the coffee is 95°F. The temperature of the water is 98°F. Which is warmer?
469) Two rational numbers are given as $-\frac{7}{9}$ and $2\frac{7}{9}$. What is the greatest absolute value?	470) A bank emails customers telling them their balance is $-50. What is the absolute value of -$50?

2.7D - Absolute Value Word Problems

EXAMPLE: The following table represents the account balance of Amanda's bank account. Which bank account has the highest absolute value?

$420.19	$1,527.09	-$2,430.49	-$3,340.40
Savings	Retirement	Checking	Other

Solution: The highest absolute value amongst $420.19, 1,527.09, -$2,430.49 and -$3,340.40 is 3,340.40. Thus, other bank accounts have the highest absolute value.

Answer: **Other**

EXAMPLE: A computer has a value of $590.99 this month. It was previously $709.99. If you want to buy a computer this month, how much money is less needed to buy a computer this month?

Solution: Subtract $490.99 from $709.99.
709.99 - 590.99 = $119. Thus, you can buy a computer for $119 less money this month.

Answer: **$119**

EXAMPLE: Bekah's weight changed by -2.9 kg last month. Amanda's weight changed by -8.1 kg last month. Who lost more weight?

Solution: The higher absolute value between -2.9 and -8.1 is 8.1. Thus, Amanda lost the most weight this month.

Answer: **Amanda**

2.7D - Absolute Value Word Problems

471) Cynthia has a balance of -$40.78. Cecilia has a balance of -$89.05. Who has the highest balance?

472) The table shows the worth of each household in the city. Which has the highest absolute value?

City Name	A	B	C	D
Price	$3,410	-$5,040	$2,310	-$2,000

473) The following table shows temperature (in degree Fahrenheit) in different cities. Which is the city that has the highest absolute value?

City Name	New Jersey	Bozeman	Newark	Macomb
Temperature	-42	-15	11	34

474) The following table shows temperature (in degree Fahrenheit) in different cities. Write temperature in order (lowest to highest absolute value).

City Name	New Jersey	Bozeman	Newark	Macomb
Temperature	-42	-15	11	34

475) The table below shows the scoreboard from a friendly game of badminton. Sort the players from the lowest to the highest absolute value of the score.

Score	-34	20	0	-7
Player	David	Nathalia	Samantha	Bella

2.7D - Absolute Value Word Problems

476) The following table represents the account balance of Cyndi's bank account. Which bank account has the highest absolute value?

$1,420.19	$527.09	-$2,130.49	-$340.40
Savings	Retirement	Checking	Other

477) Bob's weight changed by -3.9 kg last month. Amanda's weight changed by -7.1 kg last month. Who lost the most weight?

478) Bonnie is in $40 of debt. Sydney has $90.00 of debt. Cecilia has a positive bank balance of $4,010.00. Who owes the most money?

479) A computer has a value of $490.99 this month. It was previously $709.99. If you want to buy a computer this month, how much money is less needed to buy a computer this month?

480) A shoe company will have been giving discounts for the following year. It is represented in the table below. Which season has the greatest discount?

Discount	-$20	-$10	-$6.5
Season	Fall	Spring	Summer

2.7D - Absolute Value Word Problems

481) The weight changed for dogs are described in the table below. Which dog has changed the most?

Weight change	-1.9 kg	-0.45 kg
Name of the Dog	Puppy	Lola

482) The temperature of the different mixtures are given below. Which mixture's temperature is closest to zero?

Mixture	Iodine mixture	Ammonia mixture
Temperature (celcius)	5	-3

483) The following table shows the elevations of a dolphin and a sea turtle at 10:00 A.M. Which animal is farther from the sea level?

Animal	Dolphin	Sea turtle
Elevation (ft.)	31	-12

484) The price changes of a brand new phone's and a used phone's prices (in dollar amount) are given below. Which one has the greater change in price?

New Phone	-$91
Used Phone	-$32

485) The distribution of the temperature (in degree Celsius) of different cities are given below. Which city's temperature is closest to zero?

Temperature	-11	-5	4
City	A	B	C

2.7D - Absolute Value Word Problems

---

486) The distribution of the temperature of different cities are given below. Which city's temperature is farthest from zero?

Temperature	-11	-5	4
City	A	B	C

---

487) The distribution of the temperature (in degree Celsius) of different cities are given below. Write the temperature in order (lowest absolute value to highest absolute value)

Temperature	-23.15	-5.65	0
City	A	B	C

---

488) The price changes of a board game's and a toy's prices (in dollar amount) are given below. Which one has the greater change in price?

Boardgame	-$9.19
Toy	-$5.4

---

489) The price changes of a pair of shoes' and a cardigan's prices (in dollar amount) are given below. Which one has the greater change in price?

Shoe	-$19.1
Cardigan	-$13.2

---

490) The price changes of a sweater's and a sweatshirt's prices (in dollar amount) are given below. Which one has the greater change in price?

Sweater	-$11
Sweatshirt	-$2.5

2.8 - Graphing Word Problems

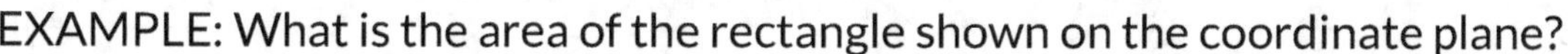

EXAMPLE: What is the area of the rectangle shown on the coordinate plane?

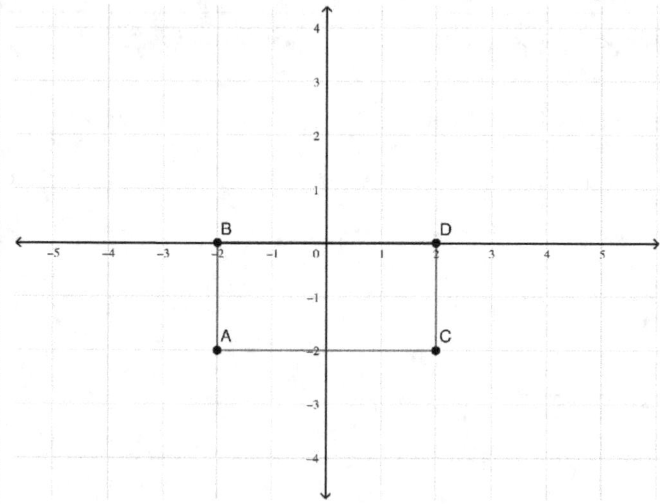

Solution: The length of AB and AC are 2 units and 4 units, respectively. If area = length × width, thus, the area of ABCD = 4 units × 2 units = 8 square units.

Answer: **8 square units**

EXAMPLE: Point A is a reflection of Point C across which axis?

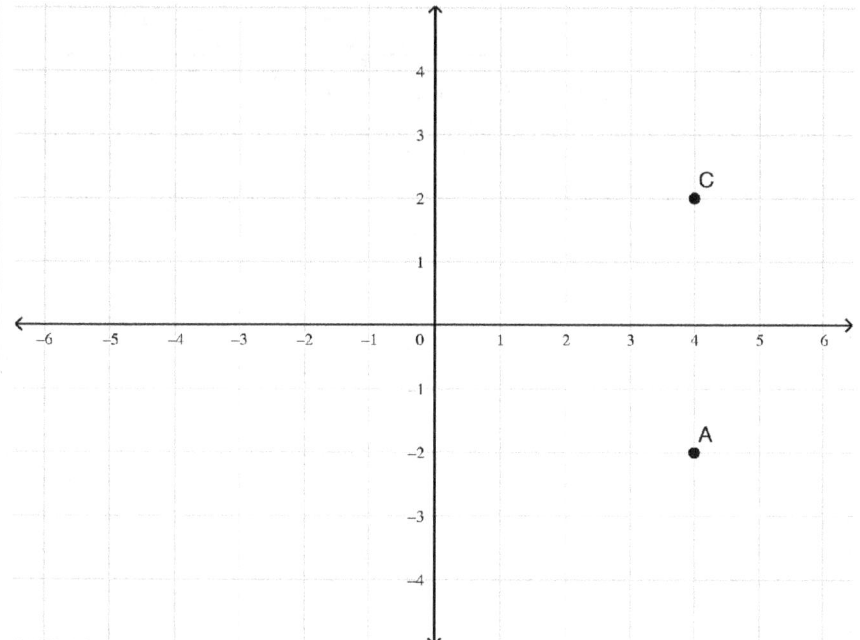

Solution: The y-axis is vertical, while the x-axis is horizontal.

Answer: **y-axis**

2.8 - Graphing Word Problems

491) To complete the rhombus, what are the coordinates for the fourth point?

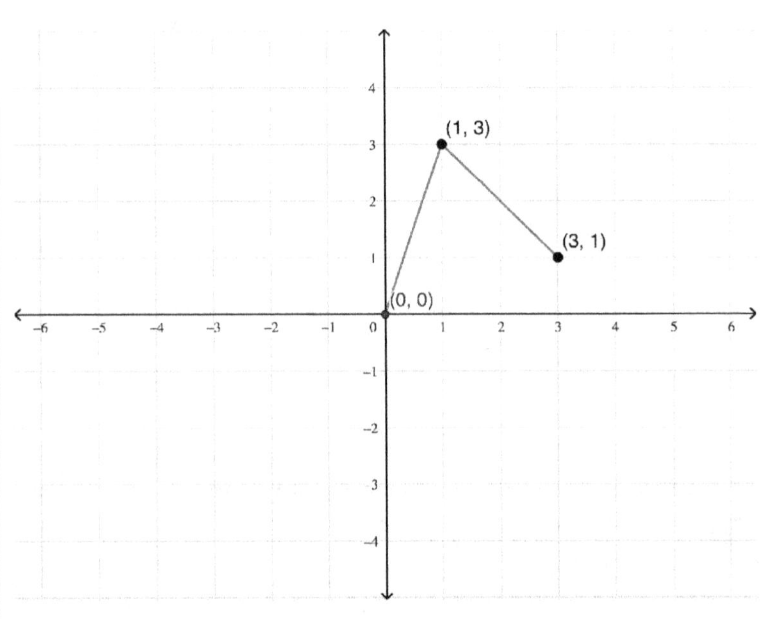

492) To complete the square, what are the coordinates for the last point?

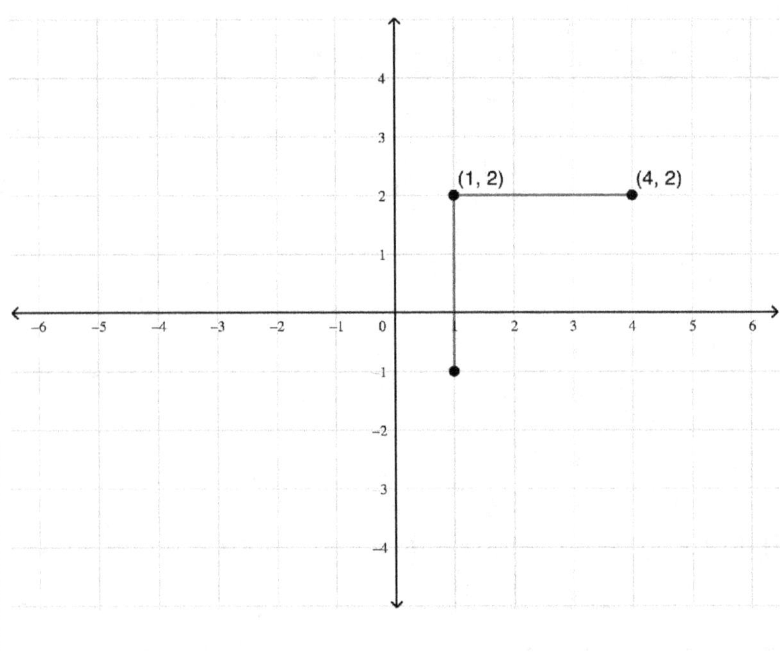

2.8 - Graphing Word Problems

493) Name the coordinates of the following polygon.

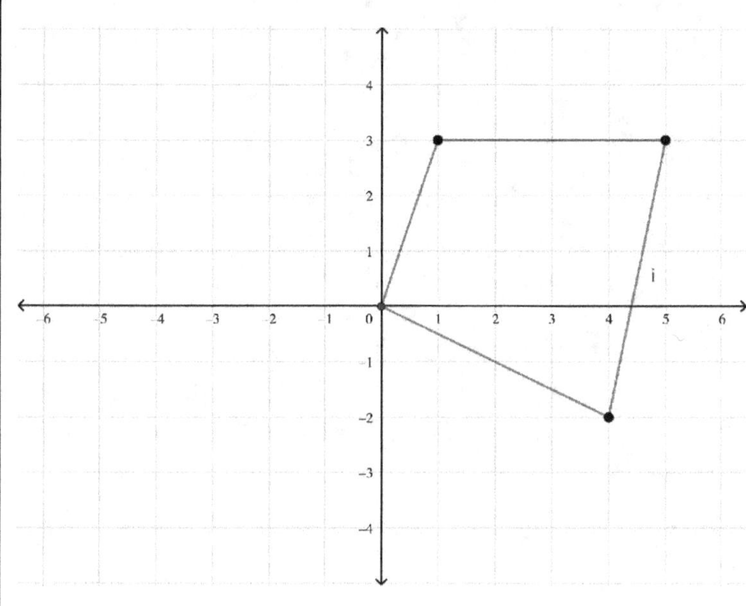

494) Point A is a reflection of Point C across which axis?

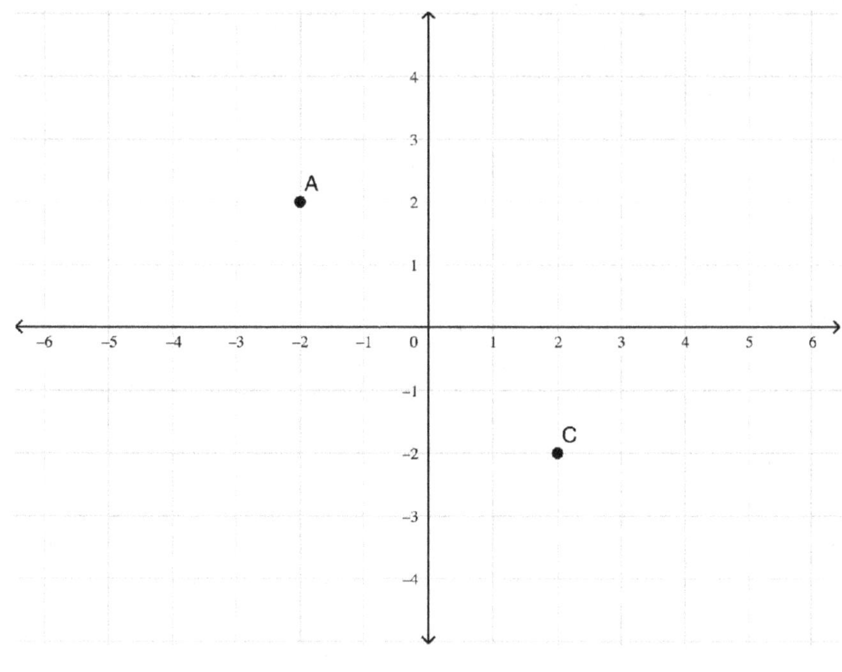

2.8 - Graphing Word Problems

495) Point A is a reflection of Point C across which axis?

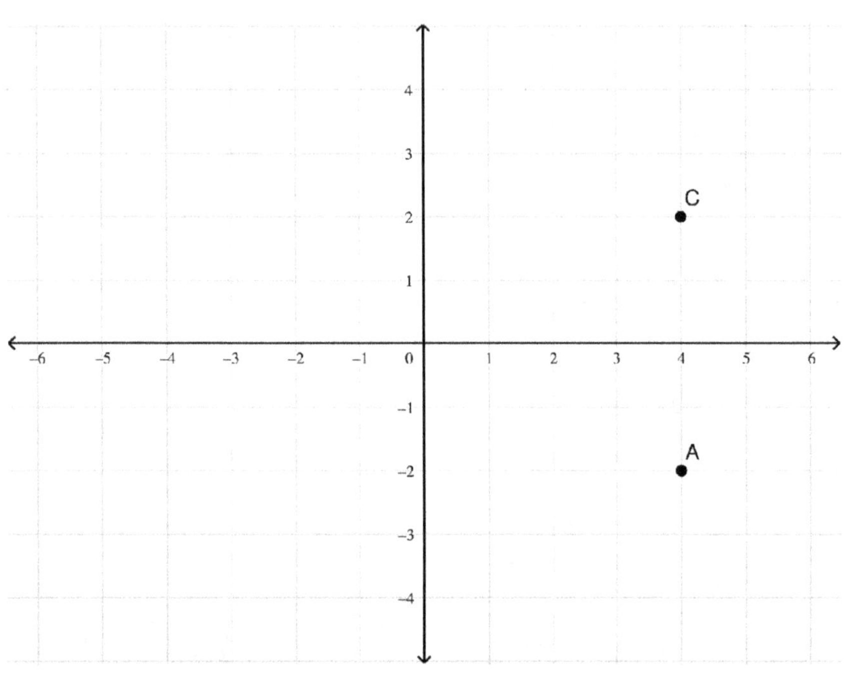

496) Point C is a reflection of Point A across which axis?

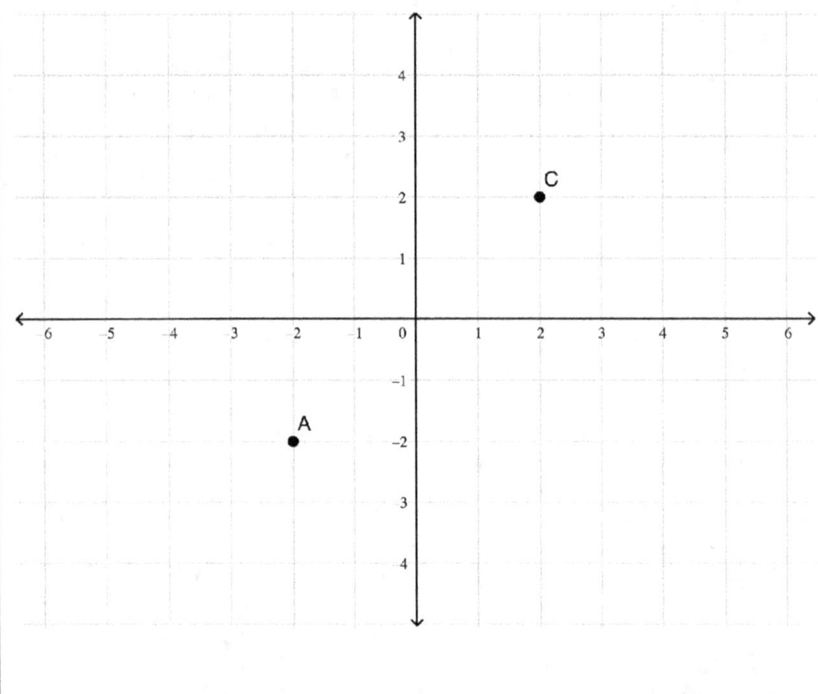

2.8 - Graphing Word Problems

497) Locate the point that represents the reflection of Point A across the y-axis.

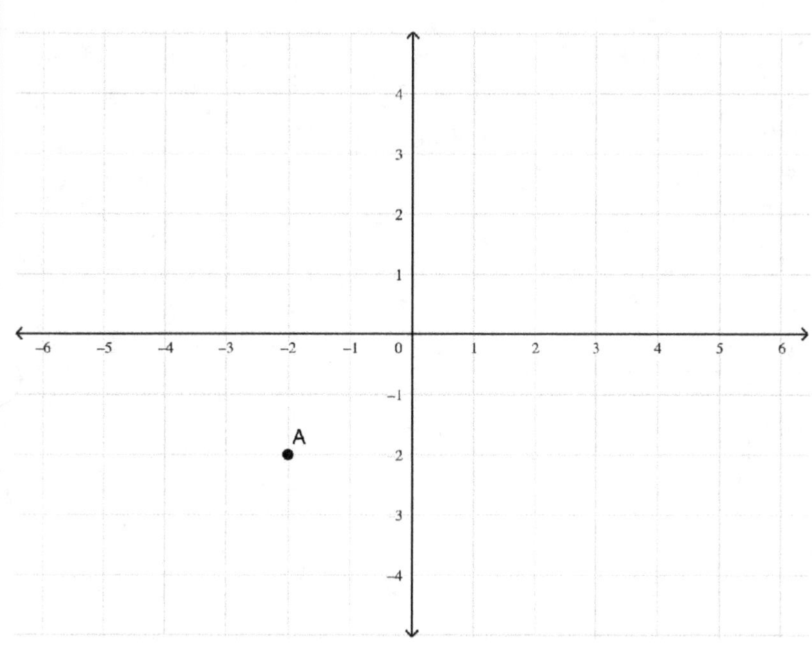

498) Locate the point that represents the reflection of Point A across the x-axis.

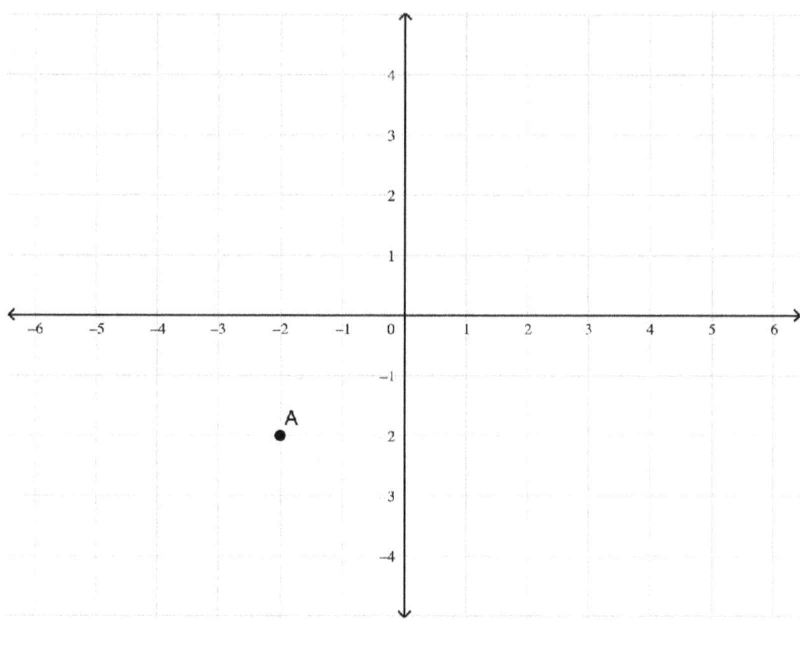

2.8 - Graphing Word Problems

499) Locate the point that represents the reflection of Point A across the x-axis, then the y-axis.

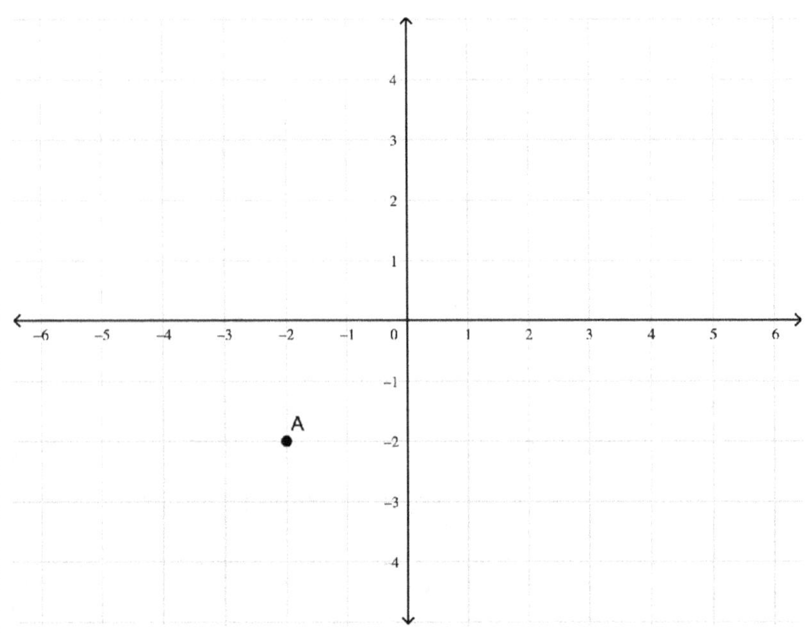

500) Find the distance between Point A and PointB.

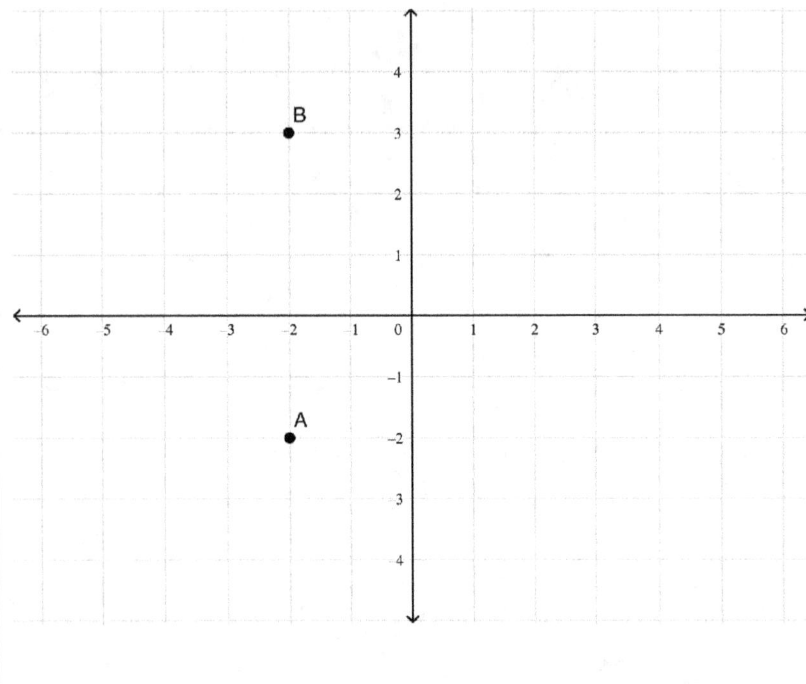

2.8 - Graphing Word Problems

501) What is the area of the rectangle shown on the coordinate plane?

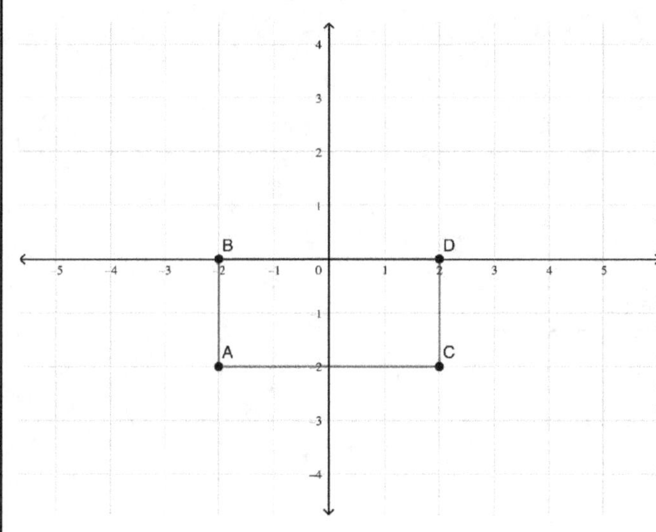

502) Are (-2, 2) and (-2, 2) located in the same quadrant?

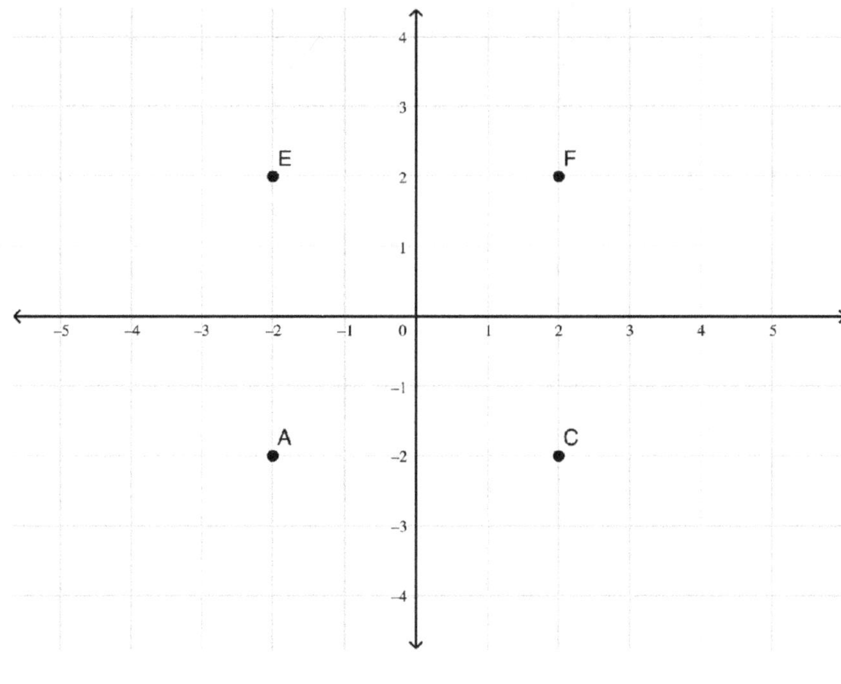

2.8 - Graphing Word Problems

503) Are (-2, 2) and (-2, -2) located in the same quadrant?

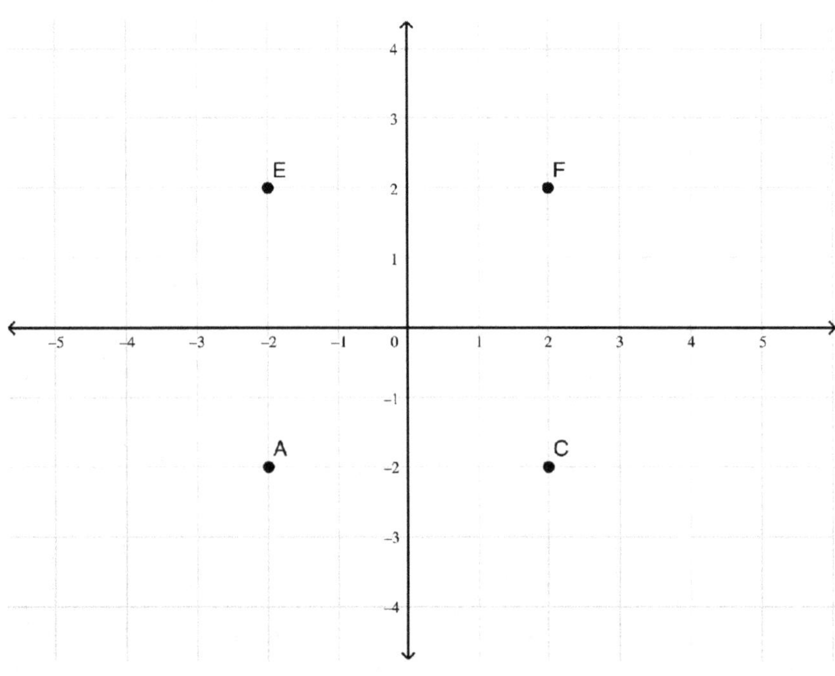

504) Are (-2, 2) and (2, -2) located in the same quadrant?

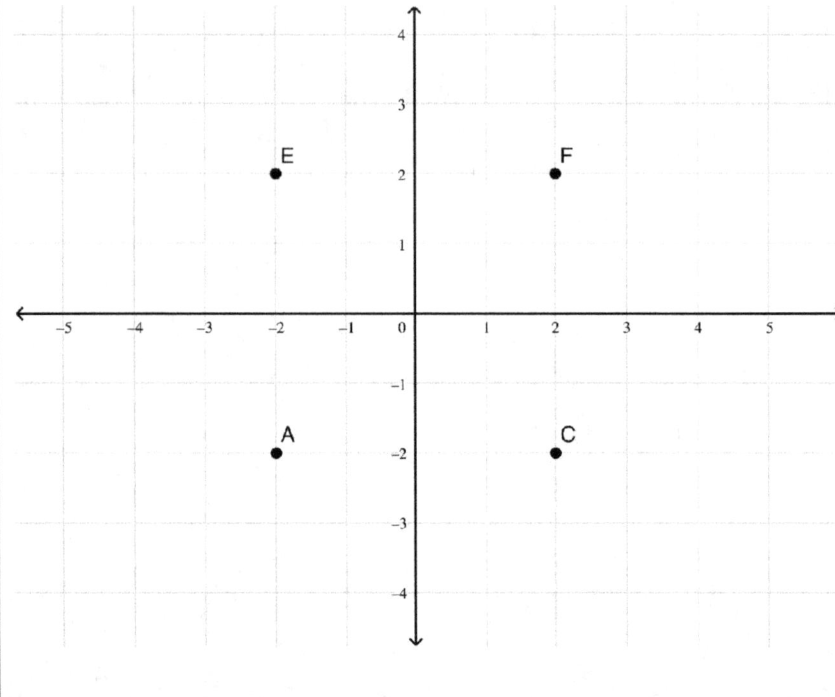

2.8 - Graphing Word Problems

---

505) What is the ordered pair of Point A?

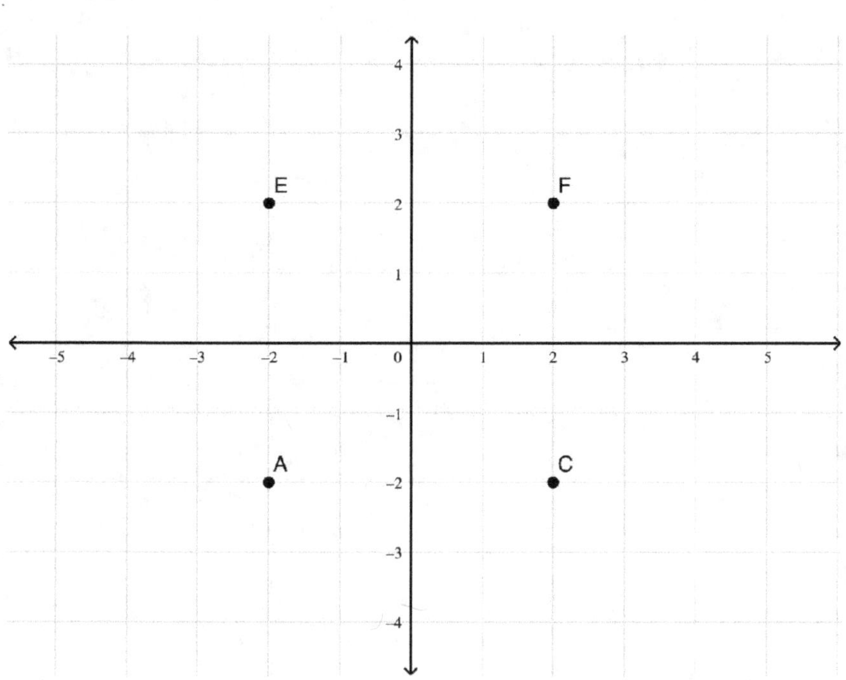

---

506) What is the ordered pair of Point E?

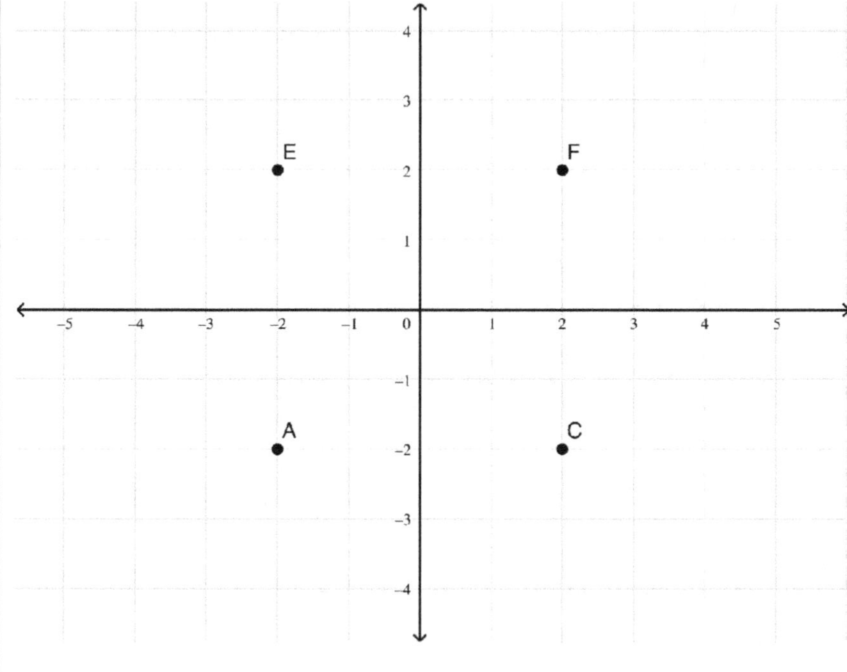

2.8 - Graphing Word Problems

507) What is the ordered pair of Point C?

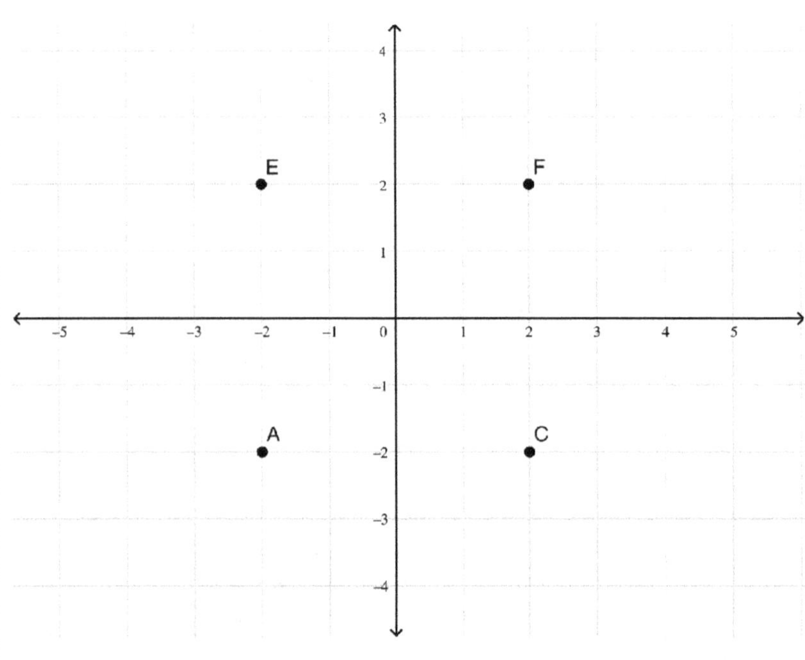

508) What is the ordered pair of Point F?

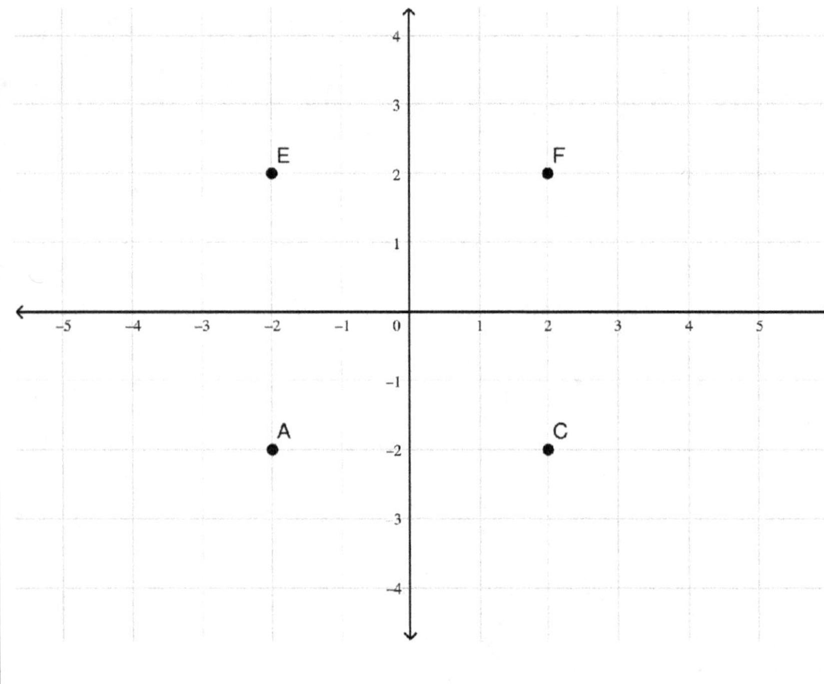

2.8 - Graphing Word Problems

509) Find the distance between Point A and Point C.

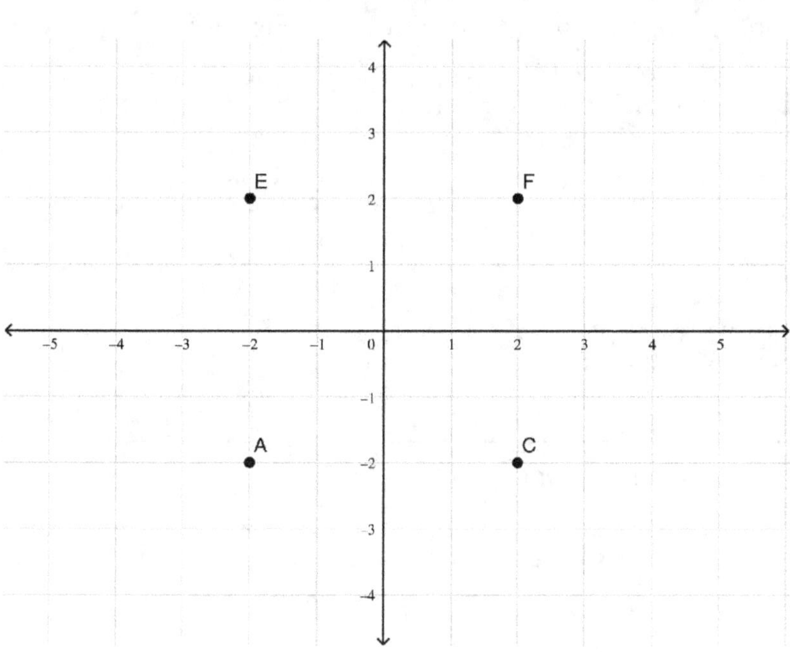

510) What is the distance between Point F and Point C?

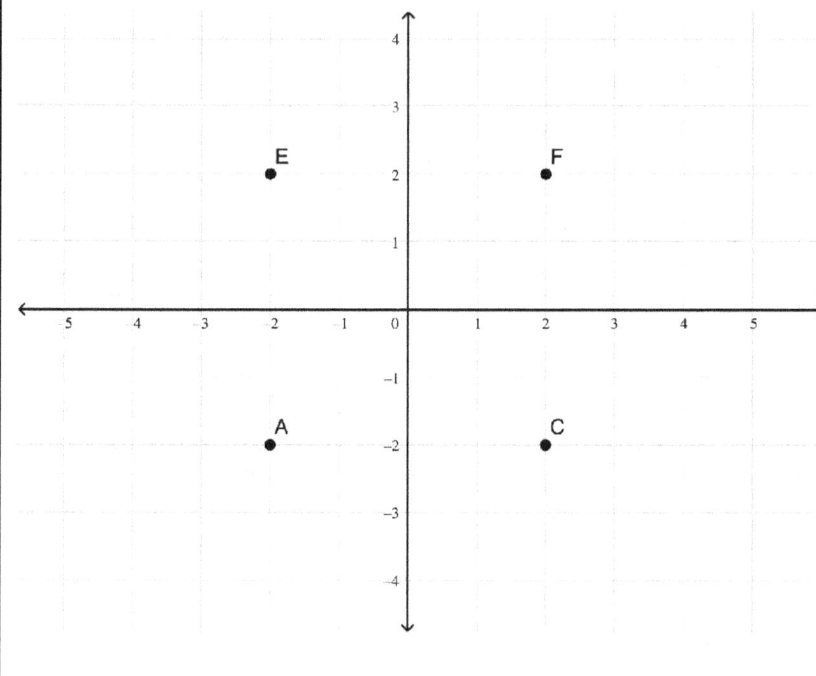

3.1 - Exponents and Basic Number Properties

An *exponent* tells how many times the base number should be multiplied by itself. There are four basic number properties.

1. *Commutative property.* For addition, a + b = b + a. For multiplication, a × b = b × a.
2. *Associative property.* For addition, (a + b) + c = a + (b + c). For multiplication, (a × b) × c = a × (b × c).
3. *Identity property.* For addition, a + 0 = a or 0 + a = a. For multiplication, a × 1 = a or 1 × a = a.
4. *Distributive property.* a × (b + c) = (a × b) + (a × c)

EXAMPLE: Is a + 4 = 4 + a true?

Solution: According to the commutative property of addition, a + b = b + a. Hence, a + 4 = 4 + a is true according to this rule.
Answer: **Yes**

EXAMPLE: Find the correct values for each exponent when $c = 5$.

Exponent	Values
(a) c^1	(i) 5
(b) 1^{5c}	(ii) 100
(c) $(2c)^2$	(iii) 1

Answer:

(a) $c^1 = 5^1 = 5$ so, (a) = (i)

(b) $1^{5c} = 1^{5\times5} = 1^{25} = 1$ so, (b) = (iii)

(c) $(2c)^2 = (2\times5)^2 = 10^2 = 10\times10 = 100$ so, (c) = (ii)

EXAMPLE: Evaluate $a^2b^3c^2$, when $a = 4$, $b = 3, c = 2$.

Solution:

We have a = 4, b = 3, and c = 2. Now, we plug these values in the expression.

Thus, $a^2b^3c^2 = 4^23^32^2 = (4 \times 4) \times (3 \times 3 \times 3) \times (2 \times 2) = 1,728$

Answer: **1,728**

3.1 - Exponents and Basic Number Properties

511) The equation $a + b + c = c + b + a$ is true according to __________ property of addition.	512) Which property of addition does hold for the following equation? $$a + b = b + a$$
513) 9 + 2 gives the same answer as 2 + 9. Is it true or false?	514) $a + 0 = a$ is true by which number property?
515) If you add zero to any number, it does not change the number that is m + 0 = m (where m is any number). What number property is it?	516) Is the equation $x + 0 = x$ valid?
517) Is the following equation true? $$xy + 0 = 0$$	518) According to the commutative property of multiplication, what is the equivalent expression for $a \times b$?
519) Does 2×3 give the same result as 3×2?	520) $x \times y = y \times x$ is true by which number property?

3.1 - Exponents and Basic Number Properties

521) Can you write $2 \times a = a \times 2$?	522) Is it true? $4 \times 5 = 5 \times 4$
523) Does the equation $p \times q = q + p$ show the commutative property of multiplication?	524) $a \times 0 = 0$ is true by which number property?
525) Which property of multiplication would always give a zero product of any number?	526) Is it true? $b \times 0 = 0$
527) Can you write $a \times 0 = 0 \times b$? [a and b any numbers]	528) Can you write $a + 0 = 0 + b$? [a and b any numbers]
529) Is the equation $x \times 0 = x$ correct?	530) Is the equation $z \times 0 = 0$ correct?

3.1 - Exponents and Basic Number Properties

531) Which property of multiplication would give the same result as the number itself after multiplication?	532) $2 \times 1 = 1 \times 2 = 2$ is true by which number property?
533) Is $2 \times 1 = 1 \times 2$ correct?	534) Here for 5^7, 5 is multiplied by itself _______ times.
535) Convert the exponential notation 10^7 to standard notation.	536) $78 + (5 + 2 - 8) \times 4 \div 2 = $ _______?
537) Find the relationship between 3^7 and 7^3.	538) If each of the four students gets four books, how many books will they get in total? Find the result as exponential notation.
539) Is the following equation true? $9 \times 9 \times 9 \times 9 \times 9 \times 9 = 9^6$	540) The prime factorization of 392 is-

3.1 - Exponents and Basic Number Properties

541) $(0.02)^3 =$ ________ ?	542) Complete the following table.

542)

	True or False
$9^3 = 9 \times 3$	a.
$2 \times 2 \times 7 \times 7 = 196$	b.

543) Determine the equivalent exponential form of $2\times2\times2\times2\times2$.	544) $(2/3)^3 =$ _____
545) $(1/7)^2 =$ _____	546) $(3/5)^3 =$ _____
547) Determine the equivalent exponential form of $7\times7\times7\times2\times2$.	548) Determine the equivalent exponential form of $4\times4\times4\times4\times4$.
549) Evaluate $x^2 y^1 z^3$, when $x = 2,\ y = 3,\ z = 4$.	550) $(3^3 + 7) + 4^2 \times 2 =$ _______ ?

3.2A - Write Expressions

An *expression* refers to the combination of mathematical symbols, numbers, and operators.

EXAMPLE: Write an expression for the following operation: multiply 5 by 9, add 24 to the product, and subtract 8.

Solution:

First, multiply 5 by 9 $\rightarrow$ 5×9

Then, add 24 to the product $\rightarrow$ $5 \times 9 + 24$

Finally, subtract 8 $\rightarrow$ $5 \times 9 + 24 - 8$

Hence, the expression is $5 \times 9 + 24 - 8$.

Answer: **$5 \times 9 + 24 - 8$**

EXAMPLE: Suppose Xian had x books and gave away his friend three books. How many books does he have now?

Solution:

Let x be the unknown number of books by Xian.

After he gave away three books to his friend, three books were less than the books' previous collection. Now, Xian has x - 3 books.

Answer: **x - 3 books**

EXAMPLE: Write an expression for the following operation: subtract 3 from the quotient of 65 and 8.

The quotient of 65 and 8 is $\frac{65}{8}$.

Now, subtract 3 from the quotient $\frac{65}{8} - 3$.

Answer: **65/8 - 3**

3.2A - Write Expressions

551) Find the value of x, when $5x = 25$.	552) Find the value of y, when $2^y + 5^y = 29$.
553) Solve $4 \times n^2 + n = 39$ for n, when $n > 0$.	554) Simplify the following expression: $2 - 2z + 6 - 8z + 4 + 3z$.
555) Julia bought x number of flowers. She gave her mother two flowers, how many flowers did she have?	556) Write an expression for the following operation: multiply two by 3, add 13 to the product, and subtract 4.
557) Write an expression for the following operation: add 7 to 9.	558) Write an expression for "y plus 4".
559) What is the expression for "y divided by 9"?	560) The expression for "7 raised to the 9th power" is-

3.2A - Write Expressions

561) Write an expression for the following operation: g is divided by 4, then multiplied by 7.	562) Eight raised to the 4th power, then the result is subtracted from y, can be written as ___________.
563) If you multiply x by y, then divide by z, what would be the final result?	564) An expression for the sum of two integers 3 and 9 is _________.
565) Subtract four from a, then take the product of the previous with z.	566) Write the expression for the following: divide a by the sum of 2 and b.
567) Write the expression for the following: divide t by the sum of 2 and n.	568) Taylor had 102 apples, then divided those apples among p students, write the expression for the apples that each student got.
569) Each adult ticket costs \$10, and each ticket for kids is \$y. If one family has five members, including three kids, then write an expression for the total ticket price.	570) There are three tigers in a zoo. One month later, the zoo authority added x number of tigers. A week later, one tiger was moved to another zoo. How many tigers are in the zoo now? Write an expression for this.

3.2B - Write Expressions with Variables

An *algebraic expression* is a mathematical expression that consists of numbers, variables, and operations. A *variable* is a symbol, usually a letter, that indicates an unknown value of a number. A *coefficient* refers to a numerical value placed before and used as a multiplicative factor of a variable. An example of an algebraic expression is 3x + 2y. The variables are x and y. The coefficients are 3 and 2.

EXAMPLE: What is the coefficient of the first term of $2y + 3xyz$?

Solution:
A *coefficient* is a number except for the variable of an expression. In this problem, we have two terms, 2y and 3xyz. The coefficient of 2y is 2, and the coefficient of 3xyz is 3.

Thus, the coefficient of the first term is 2.

Answer: **2**

EXAMPLE: ' 5 times a subtracted from 340' as an expression is-

Solution:
5 times a can be written as 5a. Now, 5a is subtracted by 340. Thus, we write 5a - 340.

Answer: **5a - 340**

EXAMPLE: What is the coefficient of the second term of $6y + 7xyz$?

Solution:
A *coefficient* is a number except for the variable of an expression. In this problem, we have two terms, 6y and 7xyz. The coefficient of 6y is 6, and the coefficient of 7xyz is 7.

Thus, the coefficient of the second term is 7.

Answer: **7**

3.2B - Write Expressions with Variables

571) Write the number of terms in the expression $ab + bc + ca$.	572) Determine the number of factors in the second term of $2y + 3xyz$.
573) One factor of $27x^2$ is 3x, is it true?	574) What is the coefficient of the second term of $2y + 3xyz$?
575) What is the coefficient of the first term of $x^2 + 4y$?	576) What is the coefficient of the first term of $2y + 3xyz$?
577) '3 times a subtracted from 120' as an expression is-	578) 'Add 9 to the quotient of 4 and 2', write it as an expression without any simplifications.
579) 'The quotient of 4 and x decreased by 1', write it as an expression without any simplifications.	580) 'The quotient of 120 and a product of 3 and 4', write it as an expression without any simplifications.

3.2C - Evaluate Expressions

EXAMPLE: Find the value of q for the following expression, when $p = 2$.

Solution:

$q = \frac{1}{2}p + 4$

$q = \frac{1}{2}(2) + 4$

$q = 1 + 4$

$q = 5$

Answer: **5**

EXAMPLE: Convert Celsius (C) scale to Fahrenheit (F) scale, if $C = 20°$.

Solution:

Celsius (C) scale to Fahrenheit (F) scale have the relation $F = \frac{c}{5} \times 9 + 32$.

Thus, from the above relation we have $F = \frac{20}{5} \times 9 + 32 = 4 \times 9 + 32 = 36 + 32 = 68$.

Answer: 68° **F**

EXAMPLE: Determine the value of the following expression, where $a = 2$, $b = 3$.
$$(a + b)^2 + a^3$$

Solution:

Here, putting $a = 2$, $b = 3$ on $(a + b)^2 + a^3$, we get $(2 + 3)^2 + 2^3 = 5^2 + 2^3 = 25 + 8 = 33$.

Answer: **33**

3.2C - Evaluate Expressions

581) What is the value of R if $P = 23$, where $R = 34 - 2 \times P$.	582) Find the value of x for the following expression, when $y = 2$. $$x = 2y + 8$$
583) Determine the value of s, from the expression $s = 56 + 4s\check{}$. Here, $s\check{} = 6$.	584) If for $x = 3$, $y = 68 - 3x$, what is the value of y?
585) If $y = 3 + 4x$, and $x = 7$, what is y?	586) Find the value of x for the following expression, when $y = 9$. $$x = 9y + 7$$
587) $a = 3b - 3$, $b = 6$, $a = ?$	588) Determine the value of m, from the expression $m = 23 - 2s\check{}$. Here, $s\check{} = 6$.
589) For $z = 10$, find the value of $y = 20z + 13$.	590) Find the value of p for the following expression, when $q = 1$. $$p = 24q - 1$$

3.2C - Evaluate Expressions

591) The area of the following rectangle with dimension 4 m by 7 m is -

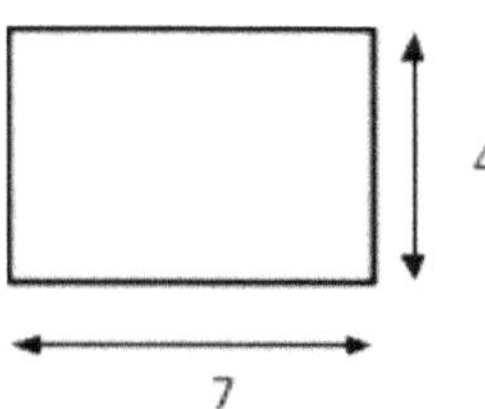

592) Determine the volume of the following rectangular box.

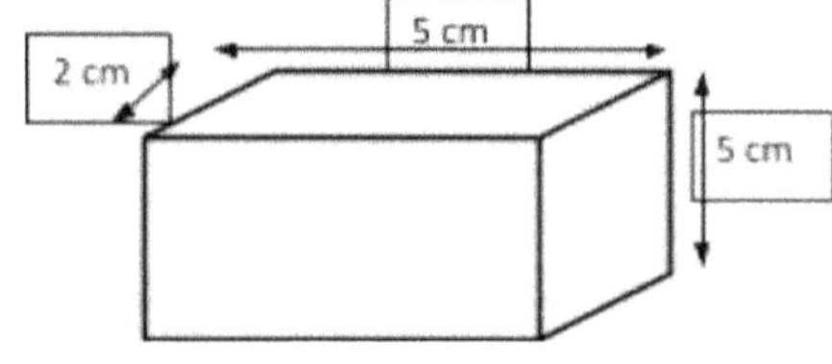

593) The area of the circle with an 8 m radius is -

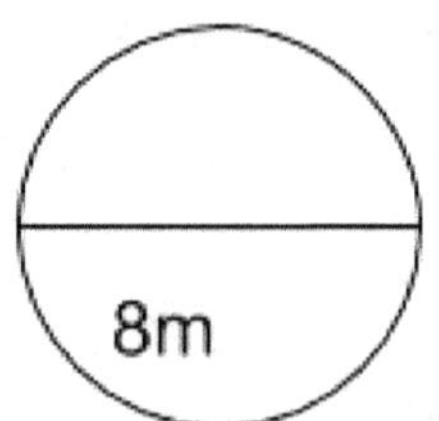

594) The area of the circle with 16 m diameter is –

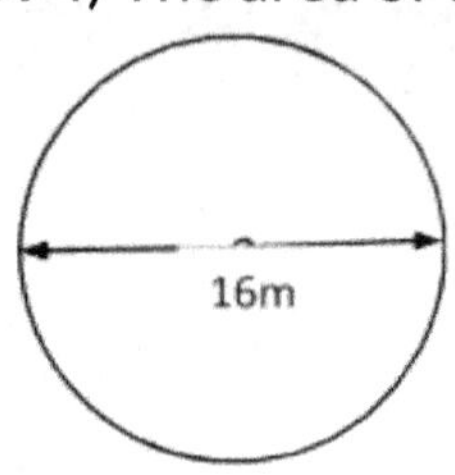

595) Consider a square with sides of 9 cm. The area of this square is-

3.2C - Evaluate Expressions

---

596) Determine the area of the following triangle.

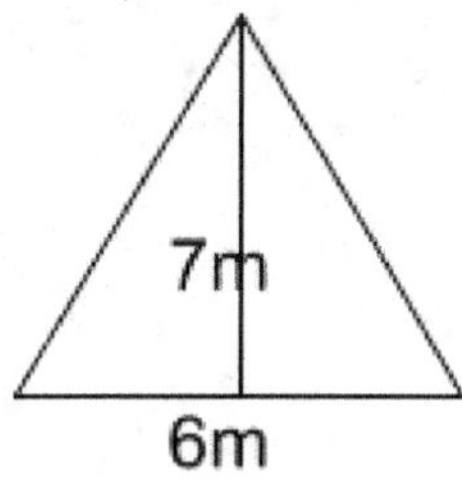

---

597) Convert Celsius (C) scale to Fahrenheit (F) scale, if they have the relation $F = \frac{C}{5} \times 9 + 32$ and $C = 40°$.

---

598) $2x + 4 - 3x =$ ______, if $x = 2$.

---

599) Applying inverse operation, find the interest rate r, if $I = \$30, \ P = \$500, \ t = 2 \ years; where \ I = P \times r \times t$.

---

600) $x - 3 = 7$, then $x =$?

3.2C - Evaluate Expressions

601) $3\times9 - 13 =$ _____?	602) $0.9 + 7.11 - 0.1\times3 =$ _______?
603) $\frac{1}{9} + \frac{5}{25} \div \frac{54}{27} =$ __________?	604) Evaluate $6x + 2 - 3,$ for $x = 9.$
605) Evaluate $6x + 2y - 3z,$ for $x = 9, y = 1, z = 2.$	606) Determine the value of the following expression, $S + \frac{5}{6},$ when $S = \frac{8}{12}.$
607) Evaluate $8x + 2 - 3,$ for $x = 2.$	608) $a\times b - 13 =$ _____, if a = 3, b = 9.
609) Ryan borrowed $200 from his sister, spent half of the money buying two textbooks, and then bought three pens for $4 each. How much does he have now? Write the related expression as well.	610) Convert Fahrenheit (F) scale to Celsius (C) scale, if they have the relation $\frac{F-32}{9}\times5 = C$ and $F = 140°.$

3.2C - Evaluate Expressions

611) If you want to increase the value of $20x$, would you increase or decrease the value of x?	612) Linda needs to increase the value of the expression $500 - 7{\times}a$. How can she do this?
613) For the changes on x, the value of the expression $\dfrac{3x^2}{2x{\times}5x}$ increases or decreases?	614) For $u = 190,\ 5u - 50 =$ _________.
615) In a school, the expression for grades' distribution is $0.3x + 0.7y$, where x stands for the grades for assignments, and y stands for the grades for tests. If Luna's grades are as follows $x = 85,\ y = 80$, what is her total grade?	616) Determine the value of the following expression, where $a = 3,\ b = 4$. $$(a + b)^2 + a^3$$
617) Determine the value of the following expression, where $p = 7,\ q = 4$. $$\frac{pq}{3} + q^4$$	618) Complete the following with proper relation: $4{\times}9^2$ _________ $12^2 + 3^3$. (Compare using, <, >, or =.)
619) Evaluate $\frac{6}{3} + 5^2 - (7{\times}4)$.	620) Evaluate $$0.5{\times}9 + \frac{4}{5}{\div}8 + (4 + 6).$$

3.3 - Equivalent Expressions and Distributive Property

EXAMPLE: Express $4y \times 5x$ into an equivalent standard form.

Solution:
$4y \times 5x = (4 \times 5)xy$ [by commutative property of multiplication, where a $\times$ b = b $\times$ a]

Thus, $4y \times 5x = (4 \times 5)xy = 20\,xy$.

Answer: **20 xy**

EXAMPLE: Factorize $6x + 4$ with the help of distributive property and taking out the greatest common factor.

Write the corresponding steps used to get the following results:

	Steps
$6x + 4$	(i) Get the GCF of the terms.
$(2 \times 3x) + (2 \times 2)$	(ii) Factor each term using the GCF.
$2 \times (3x + 2)$	(iii) Merge the common factor.

Answer: $2 \times (3x + 2)$

EXAMPLE: Simplify the expression $0.14(3 + 4x) + 2(2x + 3)$

Solution:

Step 1: By using distributive property of multiplication

$$(0.14 \times 3) + (0.14 \times 4x) + (2 \times 2x) + (2 \times 3)$$

Step 2: Multiplication, $0.42 + 0.56x + 4x + 6$

Step 3: By the commutative property of addition, add like terms.

$$0.42 + 6 + 0.56x + 4x$$

Step 4: Addition, $6.42 + 4.56x$

Answer: **6.42 + 4.56x** or **4.56x + 6.42**

3.3 - Equivalent Expressions and Distributive Property

621) Express $x \times 5y$ into an equivalent standard form.	622) Express $5 \times 2p \times 5q$ into an equivalent standard form.
623) Express $4 \times 5 \times xy$ into an equivalent standard form.	624) Express $2x \times 7y$ into an equivalent standard form.
625) Write an equivalent standard expression of the following. $$3y \times 3 \times 3x$$	626) Convert $20xy$ to an expanded expression.
627) Convert $77st$ to an expanded expression.	628) Convert $35ab$ to an expanded expression.
629) Write the corresponding steps to get the result of this expression: $3x + 3 + 4x + 5$.	630) Write the corresponding steps to get the result of this expression: $3x + 3 + (4x + 5)$.

3.3 - Equivalent Expressions and Distributive Property

631) Write the corresponding steps to get the result of this expression: $3 \times (4x + 5)$.	632) Write the corresponding steps to get the result of this expression: $(3x \times 3) \times (4x \times 5)$.
633) $4(4x + 5y) = $ _______________	634) $(x + 2y)\,5 = $ _______________
635) Factorize the expression $16x + 20y$.	636) Factorize the expression $5x + 10y$.
637) Write an equivalent expression for $2(3 + 4x) + (12x + 3)$.	638) Write an equivalent expression for $4(3 + 4x) - 2(2x + 3)$.
639) What is an equivalent expression of $3x + 9x$?	640) What is an equivalent expression of $3x + 2y + 9x$?

3.3 - Equivalent Expressions and Distributive Property

641) Simplify $12p - 7p$.	642) Simplify $7p + 13p$.
643) Use distributive property and simplify the expression $4(2x + 4z)$.	644) Use distributive property and simplify the expression $\frac{1}{2}(6x + 8z)$.
645) Using distributive property, simplify the expression $0.1(2x + 4z)$.	646) Factorize $54x + 2y$ with the help of distributive property, taking out the greatest common factor.
647) Factorize $5x + 20y + 50z$ with the help of distributive property, and taking out the greatest common factor.	648) Factorize $45 + 54y$ with the help of distributive property, and taking out the greatest common factor.
649) Is it true? $$\frac{1}{2}(3x + 10) = \frac{3}{2}x + 5$$	650) True or False? $$0.5m = \frac{m}{2}$$

3.4 - Identify Equivalent Expressions

EXAMPLE: The strip model for expression 2(x+2y) is

| x | 2y | x | 2y |

Use this to write an equivalent expression.

From the strip model, x + 2y + x + 2y = x + x + 2y + 2y (by commutative property of addition).

$$x + x + 2y + 2y = 2x + 4y$$

Answer: The equivalent expression for 2(x + 2y) is **2x + 4y**.

EXAMPLE:

Is $\frac{1}{5}(2 + x) = \frac{2}{5} + \frac{x}{5}$ true?

Start from the left-hand side, $\frac{1}{5}(2 + x)$.

Then by the distributive property, $\left(\frac{1}{5} \times 2\right) + \left(\frac{1}{5} \times x\right) = \frac{2}{5} + \frac{x}{5}$.

Hence, $\frac{1}{5}(2 + x) = \frac{2}{5} + \frac{x}{5}$ is true.

Answer: **True**

EXAMPLE: The strip model for expression 4(3x+y) is

| 3x | y | 3x | y | 3x | y | 3x | y |

Use this to write an equivalent expression.

From the strip model, 3x + y + 3x + y + 3x + y + 3x + y:

$$= 3x + 3x + 3x + 3x + y + y + y + y \text{ (by commutative property of addition)}$$
$$= 12x + 4y$$

Answer: **12x + 4y**

3.4 - Identify Equivalent Expressions

651) Is $9x + 2x + 5$ an equivalent expression of $11x + 5$?

652) Match the equivalent expressions.

(a) 4x - 2y + 1 - x	(i) 6x + 2y + 4
(b) (a + b) + (c + a) + (c + b)	(ii) 3x - 2y + 1
(c) 5x + 3y + 5 - y - 1 + x	(iii) 2a + 2b + 2c

653) Match the equivalent expressions.

(a) 3a + 4b + c + 2b + a	(i) 8a + b - 3
(b) 8a - 2 + b - 1	(ii) 4a + 6b + c
(c) (a + b) + (c + a) + (c + b)	(iii) 2a + 2b + 2c

654) Match the equivalent expressions.

(a) (3 + x) + (y - 2) + (7 - y)	(i) 16 - x + 4y - z
(b) (x + y) + (z + 2x) + (y + 3z)	(ii) 3x + 2y + 4z
(c) 2 + 6 + 8 - x + 4y - z	(iii) 8 + x

655) Is $-2 + 7x + 5$ an equivalent expression of $7x - 3$?

3.4 - Identify Equivalent Expressions

656) Find the equivalent expression of $4\left(-7 + 23x\right)$.

657) Find the equivalent expression of $5x - 3x$.

658) Write an equivalent expression of $p + 8p$.

659) The strip model for expression 2(5x + 4y) is

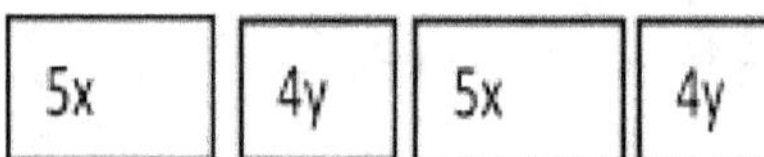

Use this to write an equivalent expression.

660) The strip model for expression 4(x + y) is

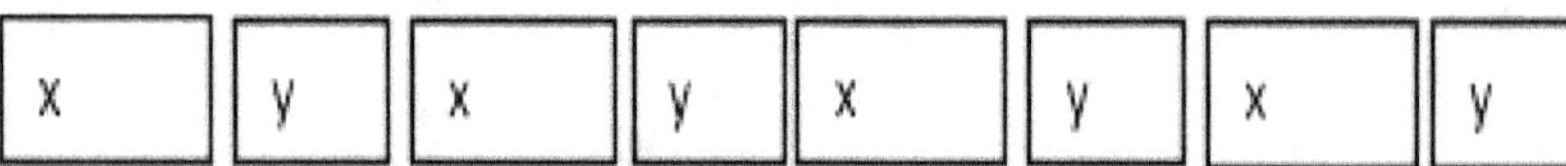

Use this to write an equivalent expression.

3.4 - Identify Equivalent Expressions

---

661) The strip model for the expression $5x + 4 + 1 + 2x$ is

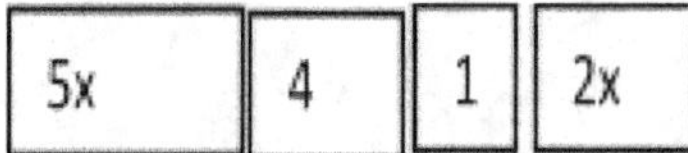

Use this to write an equivalent expression.

---

662) Simplify $p - 7p$.

---

663) Using distributive property, simplify the expression $\frac{1}{3}(9x + 27z)$.

---

664) Factorize $12 - 6x - 3z$ with the help of distributive property, taking out the greatest common factor.

---

665) Factorize (24+48) with the help of distributive property, taking out the greatest common factor.

3.4 - Identify Equivalent Expressions

666) Simplify $45 - 7$.

667) Simplify $3l + \frac{4l}{2}$.

668) Is it true? $\frac{1}{2}(2 + 10) = 1 + 5$

669) True or False

$$0.4m = \frac{2m}{4}$$

670) True or False

$$\frac{1}{5}(4 + k) = \frac{4}{5} + k$$

3.5 - Solve Equations and Inequalities

EXAMPLE: Is $x = 5$ a solution for the inequality $x \leq 7$?

Solution:
Here, $x \leq 7$, so any number equal to or less than 7 can be a solution for the given inequality.

$5 \leq 7$, thus, x = 5 is a solution.

Answer: **Yes**

EXAMPLE: Solve 4a - 6 = 10, for a.

Solution:
4a - 6 = 10   (Solve the equation step-by-step.)

4a - 6 + 6 = 10 + 6   (To extract -6, add 6 from both sides.)

4a = 16

4a/4 = 16/4   (To extract the coefficient 4, divide both sides by 4.)

a = 4

Answer: **a = 4**

EXAMPLE:

If $x = \{1, 2, 3, 5, 6, 7, 9, 8\}$, then which of the values are solutions to the inequality $x < 7$?

Solution:

From the set of values of x, we find the values which are less than 7.

The values are 1, 2, 3, 5, 6.

Answer: **1, 2, 3, 5, 6 are solutions to the inequality** $x < 7$.

3.5 - Solve Equations and Inequalities

671) Is the inequality $2x < 6$ true for $x = 1$?	672) Can $x = 2$ be a solution for the inequality $x \leq 3$?
673) Liz wants to buy two burgers. She has $5, and each burger costs $3. Can she buy two burgers?	674) Abe cannot run more than 2 miles total. In a competition he is required to run six 0.5 mile races. Can he do that?
675) Is the inequality $3 > y$ true for $y = 0.5$?	676) Does the inequality $2p \geq 12$ hold for $p = 6$?
677) In lab work, Natalie was given an equation $3x + 2 = 3.5$, where x is her experimental data. If she obtained $x = 0.5$ from her experiment, does the value of x satisfy the given equation?	678) Solve $\frac{1}{2} + \frac{x}{3} = \frac{9}{5}$ for x.
679) Solve $3x + 6 \times \frac{9}{2} = 29.4$	680) Does $y = 2$ satisfy $66 = 23y + 10$?

3.5 - Solve Equations and Inequalities

681) Solve $3a - 7 = 5$.	682) Three sides of a triangle are a, b, and c. The perimeter is 12, and $a = 3$, $c = 5$, then $b = $? .
683) Two rats were fighting over one bread; one took $\frac{3}{4}$ th of the bread. How much did the other one get?	684) Show the steps to solve the equation $2x + 4 = 10$.
685) Solve $7y = 2y - 5 + 80$.	686) Solve $y + 3 - 40 = 2y - 6 + 120$.
687) For $y = 7$, $2y - 68 + 3y + 120 = $?	688) Find the solution of $7a + 3 = -7 - 3a + 56$.
689) Find the solution of $18a = -18$.	690) What are the positive integers in $x < 5$?

3.5 - Solve Equations and Inequalities

691) What are the negative integers in $x < -5$?	692) Consider $x = \{1, 2, 5, 6, 7, 9, 8\}$, then which of the values are solutions to the inequality $x \le 6$?
693) Consider $y = \{1, 2, 5, 6, 7, 9, 8\}$, then which of the values are solutions to the inequality $x \ge 6$?	694) Is $\frac{1}{z} + 1 \le 2$ true for $z = 1$?
695) Is $\frac{1}{k} + 1 = 2$ true for $k = 1$?	696) Is $\frac{2}{k} + 3 = 4$ true for $k = 2$?
697) Is $4a + b = 15$ true for $a = 3$ and $b = 3$?	698) Consider $y = \{1, 2, 5, 6, 7, 9, 8\}$, then which of the values are solutions to the inequality $y \ge 7$?
699) Is $5ab = 20$ true for $a = 4$ and $b = 2$?	700) Consider $y = \{2, 4, 6, 8, 10, 12\}$, then which of the values are solutions to the inequality $y < 9$?

3.6 - Write Variable Expressions

EXAMPLE:

Jessica had \$20, and her father gave z more. Now, she has a total of \$55. How much did her father give her?

Solution:

Accordingly, $20 + z = 55$.

$20 + z - 20 = 55 - 20$

$z = 35$

Hence, Jessica's father gave her \$35.

Answer: **\$35**

EXAMPLE:

Solve 'a multiplied by 5 is 25' for a.

Solution: We write the equivalent expression as $5a = 25$.

$5a = 25$

$5a/5 = 25/5$

$a = 5$

Answer: **5**

EXAMPLE:

Jerry wants to buy x books. Each book costs \$2. How much does he need to buy x books?

Solution:

One book costs \$2.

Then, x books cost \$2x.

Answer: **\$2x**

3.6 - Write Variable Expressions

701) Oliver had nine apples; from those, he gave a apples to her sister. How many apples does he have now?	702) Noah bought a book for $\$x$ and spent twice on food. How much did he spend on food?
703) Harry wants to buy x sandwiches. Each sandwich costs $\$2$. How much does he need to buy x sandwiches?	704) Abella cannot run more than y miles. In competition, he is required to run at least 3 miles. If he wants to participate in the competition, write an expression for y accordingly.
705) If a triangle has three sides of length x, y, and z. Write an expression for its perimeter.	706) The length of one side of a square is x cm. If the side is increased by 2 cm, then what would be the length of each side of the later square?
707) The length of one side of a square is x cm. If the side is decreased by 3 cm, then what would be the length of each side of the later square?	708) Among x pieces of vanilla and chocolate-flavored cake, there are two pieces of chocolate-flavored cake. How many pieces of cake are vanilla-flavored?
709) Write an equation for the following, 'sum of x and 20 is 30'.	710) Solve 'sum of x and 20 is 30' for x.

3.6 - Write Variable Expressions

711) Write an equation for the following, 'a multiplied by 5 is 25'.	712) Solve 'a multiplied by 5 is 25' for a.
713) There are 20 blueberries in Harry's garden and x blueberries in Tom's garden. How many blueberries do they have in total?	714) Jessica had \$20, and her father gave z more. Now, how much money does she have?
715) Mark had a collection of 30 stamps, and her friend gave z more. Now, she has a total of 97 stamps. How many stamps did her friend give her?	716) Monica was wrapping gift packs, where 36 gifts were put in N bags. Each bag contains four gifts. Write an equation for this.
717) Monica was wrapping gift packs, where 36 gifts were put in N bags. Each bag contains four gifts. How many bags were there?	718) Johnny gives Tina 12 dollars to buy y soft drinks. Each one costs 4 dollars. Write an equation for this.
719) Johnny gives Tina 12 dollars to buy y softdrinks. Each one costs 4 dollars. How many soft drinks can Tina buy?	720) Zach buys w shirts, each for \$8. He needs \$96 in total. Write an equation for this.

3.7 - Solve One-Step Equations

EXAMPLE: Solve $z \times 4 = 2 \times 3$.

Solution:

$z \times 4 = 2 \times 3$ (Multiply both sides.)
$4z = 6$

$$\frac{4z}{4} = \frac{6}{4}$$ (Divide both sides by 4.)

$$z = \frac{3}{2}$$ (Simplify the fraction, if necessary.)

Answer: **3/2**

EXAMPLE: Solve $8q = 96$.

Solution:
$8q = 96.$

$$\frac{8q}{8} = \frac{96}{8}$$ (Divide both sides by 8.)

$q = 12$

Answer: **12**

EXAMPLE: Solve $z \div 4 = 2 \times 3$

Solution:

Here,

$z \div 4 = 2 \times 3$

$$\frac{z}{4} = 6$$

$$\frac{z}{4} \times 4 = 6 \times 4$$

$z = 24$

Answer: **24**

3.7 - Solve One-Step Equations

721) Solve $12p = 144$.

722) What is the value of y, if $y + 1 = 37$?

723) What is the value of v, if $v + 1 = 3$?

724) Solve $\frac{2x}{7} = 30$.

725) What is the value of x, if $8 = x + 2$?

3.7 - Solve One-Step Equations

726) If $k \times 7 = 2$, what is the value of k?

727) If $h \div 9 = 2$, what is the value of h?

728) Solve $p \times 4.5 = 9$

729) Solve $18q = 72$.

730) Solve $4q = 92$.

3.7 - Solve One-Step Equations

731) What is the value of w, if $6w + 1 = 37$?	732) What is the value of s, if $\frac{s}{2} = 37$?
733) Solve $x - 3 = 54$.	734) Solve $\frac{3x}{7} = 3$
735) Solve $w + \frac{2}{3} = \frac{4}{3}$.	736) Solve $z \div \frac{7}{2} = 45$.
737) Solve $3y - 0.4 = \frac{45}{2}$.	738) Solve $q \div 3 = 4 \times 6$.
739) Solve $z \times 2.5 = 9$.	740) Solve $r - 6 = 0$.

3.7 - Solve One-Step Equations

741) What is the value of x, if $3 = x - 2$?	742) If $x \div 7 = 2$, what is the value of x?
743) If $k \times 2 = 1$, what is the value of k?	744) There are nine birds sitting on a branch of a tree. Two of them are blue, and x others are black. What equation can be formed to find the number of black birds?
745) Julie picked up w decorative pieces for $25. Each piece cost her $2.5. What is the equation that can give the value of w?	746) There are x pages in a detective book. The book author prints five pages more and finally gets a book of 57 pages for a special edition. Write the appropriate equation for x.
747) Mr. William brought w notebooks for his two children. Each one got seven notebooks. Write the appropriate equation for w.	748) Find a value for x, that satisfies $x - 3 = 11$.
749) Find a value for x, that satisfies $x^2 - 1 = 8$, when $x > 0$.	750) Find a value for x^3, if $x = 4$.

3.7 - Solve One-Step Equations

751) Solve $x - 2 > 8$.	752) Solve the inequality $30y + 5 \geq 65$.
753) Solve $6x - 12 > 48$.	754) Solve the inequality $3y + 1 \geq 7$.
755) Simplify $3a + 4b = 7a - a + b$.	756) Shana is taller than Ryan. Ryan is 5 feet, and Shana is h feet tall. Write the appropriate inequality for h.
757) Find a value for x, that satisfies $$\frac{x+7}{7} + 1 = -3 + 70$$	758) Find the value of y considering the following equation: $(6 + 6 + 6) = 7y$.
759) Jolly solved the equation $3y + 1 = 7$. The steps he used for solving this: [step 1] 3y + 1 - 1 = 7 + 1 [step 2] 3y = 8 [step 3] 3y/3 = 8/3 [step 4] y=8/3 Where does she make a mistake (if any)?	760) Solve 42x = 6.

3.8 - Solve and Graph Inequalities

EXAMPLE:

Find the solution x from $3x + 3 \geq 24$.

Solution:

$3x + 3 \geq 24$

$3x + 3 - 3 \geq 24 - 3$ (Subtract both sides by 3.)

$3x \geq 21$

$3x/3 \geq 21/3$ (Divide both sides by 3.)

$x \geq 7$

Answer: **x ≥ 7**

EXAMPLE: Find the inequality for y using the number line.

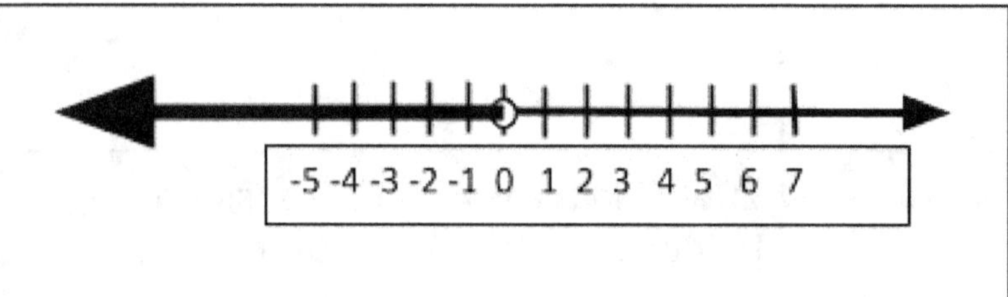

Solution:

In the number line, y takes all the negative values and can be equal to 0. The open circle indicates that 2 is NOT a solution. Thus, this can be represented as y < 0.

Answer: **y <0**

3.8 - Solve and Graph Inequalities

761) Find the inequality for x, from the following number line.

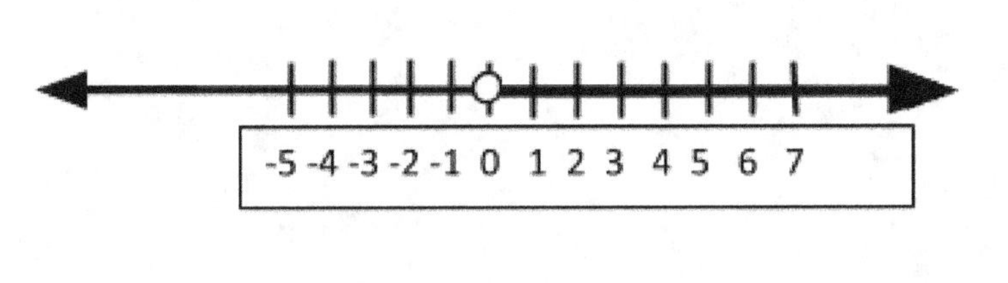

762) Find the solution x from $2x + 2 \geq 24$.

3.8 - Solve and Graph Inequalities

763) Find x from $\frac{x}{4} + 5 \leq 17$.

764) Is the following number line is a true representation for $y < 0$?

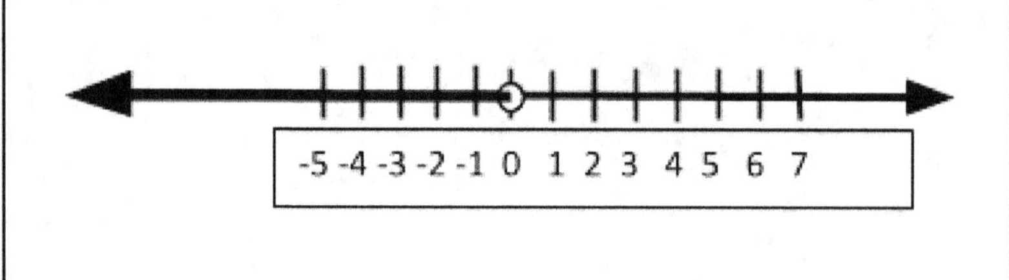

3.8 - Solve and Graph Inequalities

765) Write the inequality for y using the number line.

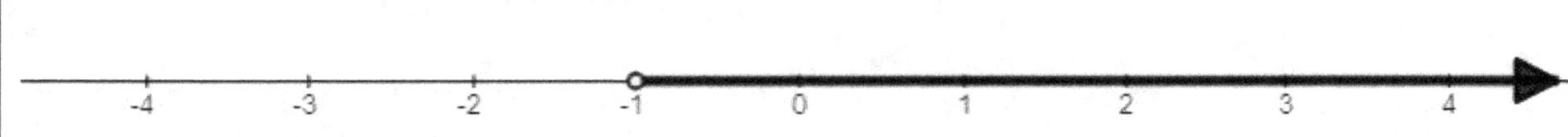

766) Tina can give her friend no more than two pencils. She gave her g pencils. Now write an inequality for g.

3.8 - Solve and Graph Inequalities

767) Determine r from $\frac{r}{3} \geq 3$

768) Use the inequality $x > -2.5,$ and place the value of x in the number line.

3.8 - Solve and Graph Inequalities

769) Graph the values of w that is derived from the inequality $2w < -2$.

770) Solve $-\frac{3x}{7} \geq 5$ and determine x.

3.9 - Dependent and Independent Variables

Dependent and independent variables are commonly used in mathematics, statistics, and science. The *independent variable* is the value that can be controlled or changed. The *dependent variable* is the value that depends on the independent variable.

EXAMPLE: In a shop, more bottle juices are sold if the temperature is high. If t represents temperature and c represents the number of bottled juices sold, identify the independent and dependent variables.

Solution: If the temperature increases, more people will buy bottled juices. However, if the temperature decreases, fewer people will buy bottled juices. Therefore the number of bottled juices sold depends on the temperature.

Answer: **The temperature t is the independent variable. The number of bottled juices sold c is the dependent variable.**

EXAMPLE: A teacher in a school gives marks as $x = y + 2$ at the end of each semester, where y represents the obtained marks, and 2 is a bonus. Identify the independent and dependent variables.

Solution: Here, the total mark x of each student depends on the obtained mark y.

Answer: **The obtained mark y is the independent variable. The total mark x is the dependent variable.**

EXAMPLE: Nora gets more money when she can sell more flowers. The following graph shows the money that Liana gets y, and the number of flowers x she can sell. Identify the independent and dependent variables.

Solution: To get more money, Nora needs to sell more flowers. Therefore, fewer flowers sold mean fewer sales. Her earnings totally depend on the number of flowers she can sell.

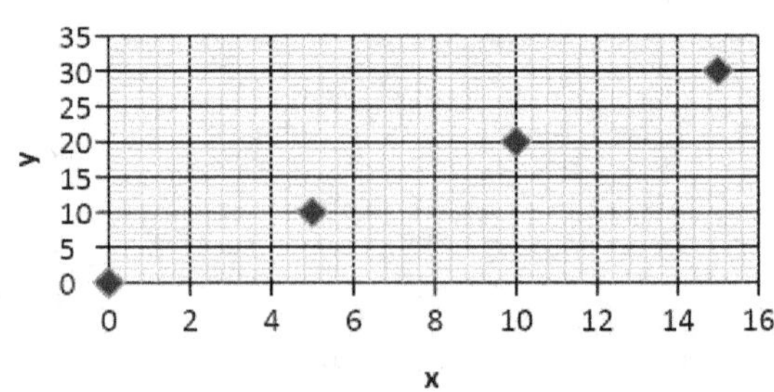

Answer: **The number of flowers sold x is the independent variable. The earning y is the dependent variable.**

3.9 - Dependent and Independent Variables

771) Tina and Liana get candy from their mother whenever they help her in making cookies. The mother gives more candy to whoever makes more cookies. If a represents the number of candies and b represents the number of cookies, identify the independent and dependent variables.

772) The teacher gives more marks to the students in class work who can solve more class problems. If x represents the marks that a student gets and y represents the number of problems a student can solve, identify the independent and dependent variables.

773) An ice cream seller sells more ice creams if the temperature is high. If t represents the temperature and c represents the number of ice creams that he can sell, identify the independent and dependent variables.

774) Robbie can play more whenever he has fewer classes to attend in school. If t represents the time for playing and c represents the number of classes, identify the independent and dependent variables.

775) Don has to read more whenever he has lots of classes to attend in school. If t represents the study time and c represents the number of classes, identify the independent and dependent variables.

3.9 - Dependent and Independent Variables

776) Lily gets more marks in assignments when she can solve more problems. The following table shows the marks that Liana gets, x, and the number of problems, y, she can solve.

x	0	5	10	15
y	0	10	20	30

Identify the independent and dependent variables.

777) Solve $8x = 3x + 10$.

778) Kylie buys more peppers when she gets a few peppers from her garden.
If S represents the number of peppers from the shop and G represents the number of capsicums from her garden, identify the independent and dependent variables.

779) Solve $4p = 12q - 8,\ when\ q = 3$.

780) Barry buys more capsicums when she gets a few capsicums from her garden. The following table shows the number of capsicums that she buys from shop S and the number of capsicums she gets from the garden, G.

S	25	21	15	0
G	0	4	10	25

Identify the independent and dependent variables.

3.9 - Dependent and Independent Variables

781) What is the value of x , when given $x = 4y + 3$ and $y = 1$.

782) Solve $4p = 3q - 7$, when $q = 5$.

783) The age of the son is $x = y - 25$, when his father is $y = 45$ years old. How old is the son?

784) In a flower shop, the total profit is $z = 10f - 5$, where f is the number of sold flowers. If $f = 78$, then $z = $?

785) The English teacher gives marks as $x = y + 5$ at the end of each semester. Where y the obtained marks that a student gets and 5 is a bonus. Complete the following table

Y	20	15	25
X	25	20	

3.9 - Dependent and Independent Variables

786) A farmer gets profit $x = 3y + 100$, where y is the amount he invests. Complete the following table.

y($)	1,000	1,500	1,700
x($)	3,100		5,200

787) A farmer gets profit $x = 3y + 100$, where y is the amount he invests. Complete the following table.

y($)		700	900
x($)	1,600	2,200	2,800

788) In a flower shop, the total profit is $z = 6f + 115$ where f is the number of sold flowers. Complete the following table.

z	295	415	535
f	30		70

789) Draw a graph using the equation $y = x$.

790) Draw a graph using the equation $y = 0.3x$.

3.9 - Dependent and Independent Variables

791) The following graph shows the relationship between shirts and its price ($). Along the horizontal axis the number of shirts and along with the y-axis total price of the shirts.

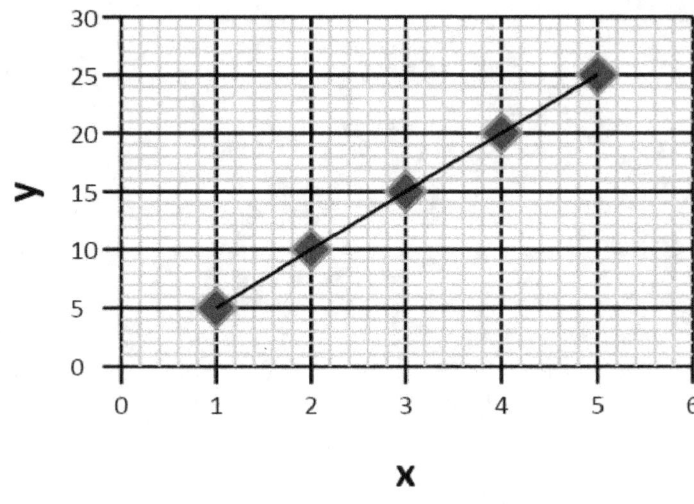

What is the price of 5 shirts?

792) Lara has $x more than Tammy. If Tammy has x then Lara has y. Now, from the following graph, how much more money does Lara have than Tammy at all times?

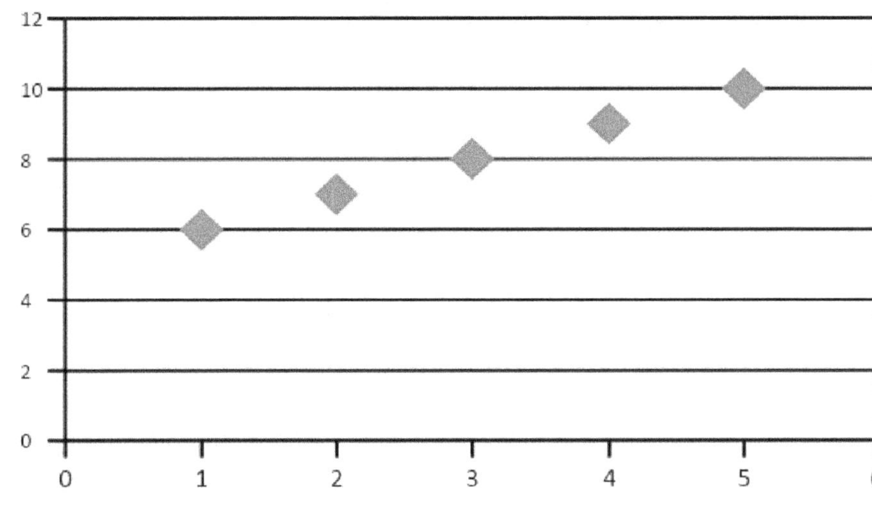

3.9 - Dependent and Independent Variables

793) Complete the table using the equation $y = 2x + 1$.

x	y
(a) 2	5
(b) 4	
(c)	15

794) Draw a graph using the equation $y = 2x + 1$.

3.9 - Dependent and Independent Variables

795) Complete the table using the equation $y = x - 3$.

x	y
(a) 10	7
(b) 15	
(c)	20

796) Draw a graph using the equation $y = x - 3$.

3.9 - Dependent and Independent Variables

797) Write an equation $y =$ _______ that gives the values in the following table.

x	y
3	4
9	10
17	18

798) Write an equation for w that gives the values in the following table.

z	w
5	15
9	27
12	36

3.9 - Dependent and Independent Variables

799) Using the graph, complete the table and write the equation involving the variables.

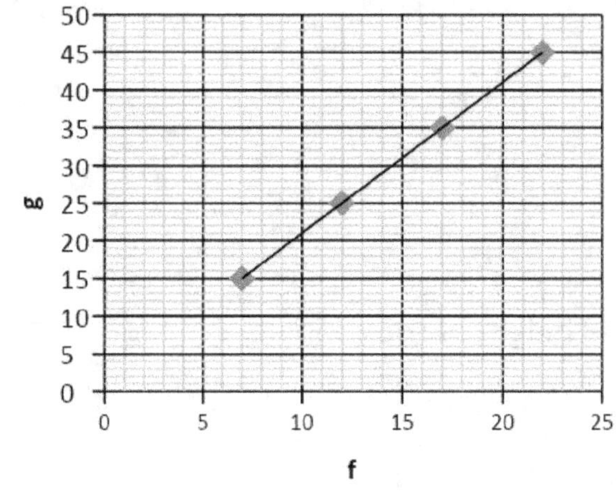

f	7	12	17	22
g	15		35	45

800) Using the graph, complete the table and write the equation involving the variables.

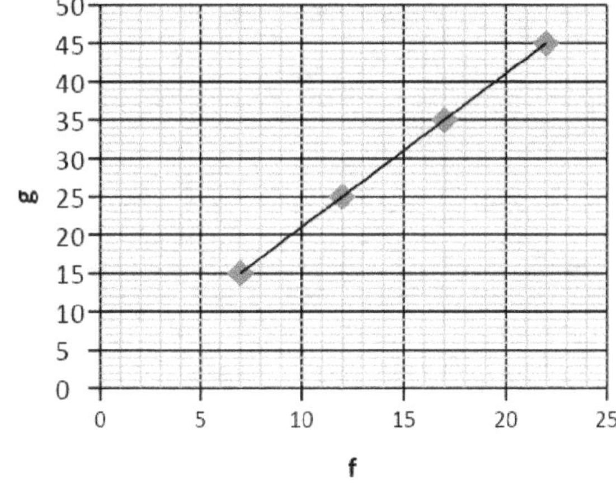

f	4	8	12	16
g	12	24	36	

4.1 - Find the Area

EXAMPLE: Calculate the area of ABC

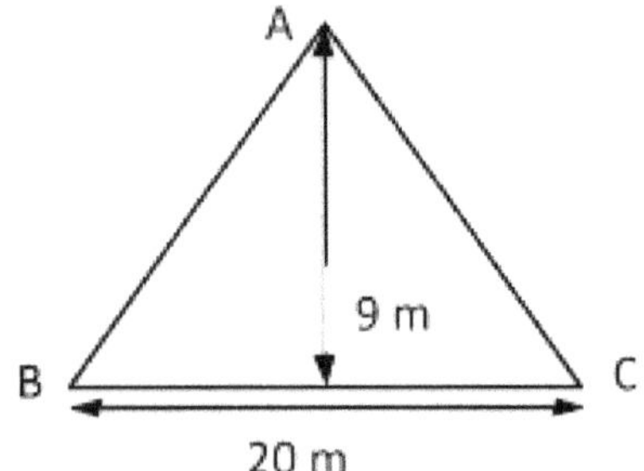

To find the area of a triangle, we need the base and height of the triangle.

The base of ABC, BC = 20 m

The height of ABC = 9 m

Answer: The area of ABC is = $\frac{1}{2}$ (base × height) = $\frac{1}{2}$ (20 m × 9 m) = $\frac{1}{2}$ (180 m)

= **90 square meters**

EXAMPLE: Calculate the area of the compound figure.

This figure includes one rectangle and
one triangle. To calculate the area of the
whole shape, we need to determine the
area of the triangle and the rectangle
individually and then add them.

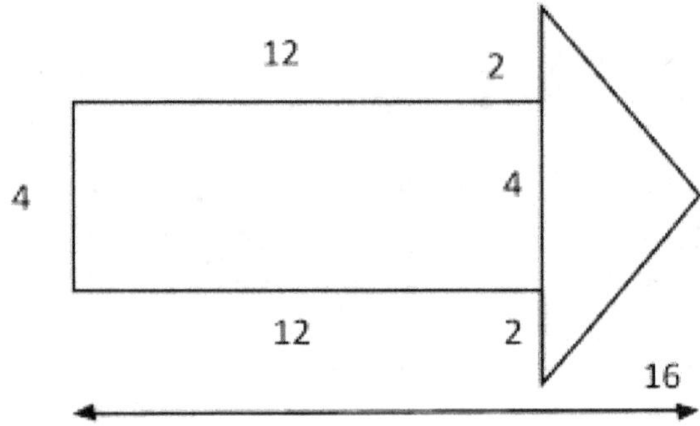

Step 1 :

The area of triangle = $\frac{1}{2}$ (base × height)

Here, base = 4 + 2 + 2 = 8

Height = 16 - 12 = 4

Hence, the area of triangle = $\frac{1}{2}$ (8 × 4) = 16

Step 2: The area of the rectangle = (length × width)

Length = 12

Width = 4

The area of the rectangle = (length × width) = 12 × 4 = 48

Answer: The area of the compound figure is 16 + 48 = **64 squares units.**

4.1 - Find the Area

---

801) Calculate the area of ABC.

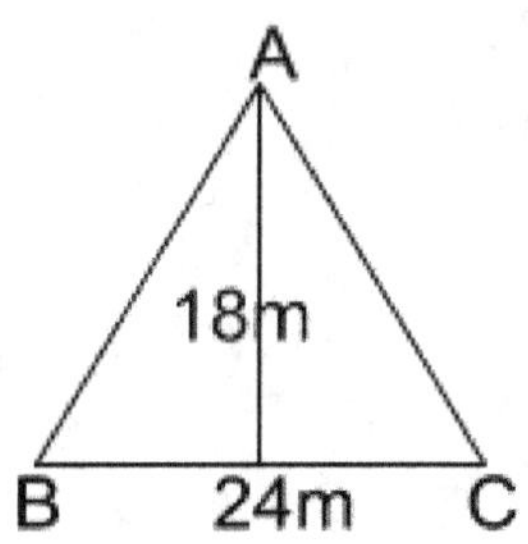

---

802) Calculate the area of ABC.

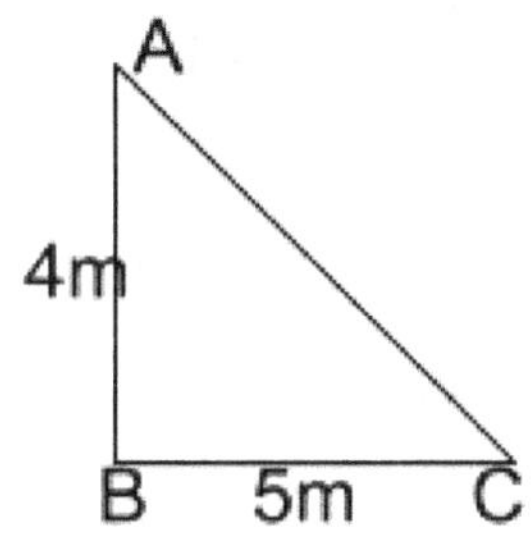

4.1 - Find the Area

803) Michelle has a rectangular object with $7\ cm$ length and $5\ cm$ width. Find the area of the rectangle.

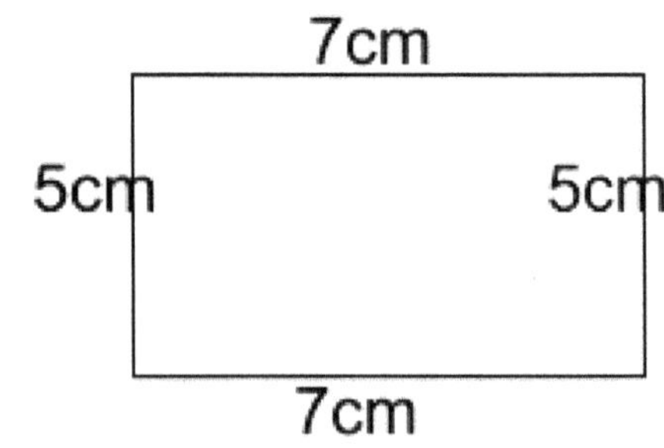

804) A triangle has an area of $35\ m^2$. The base BC of the triangle is $14\ m$. Determine the height AD of the Triangle ABC.

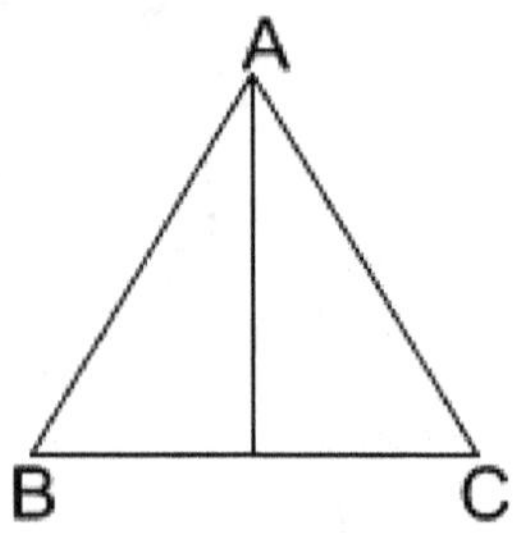

4.1 - Find the Area

805) Determine the area of the following triangle.

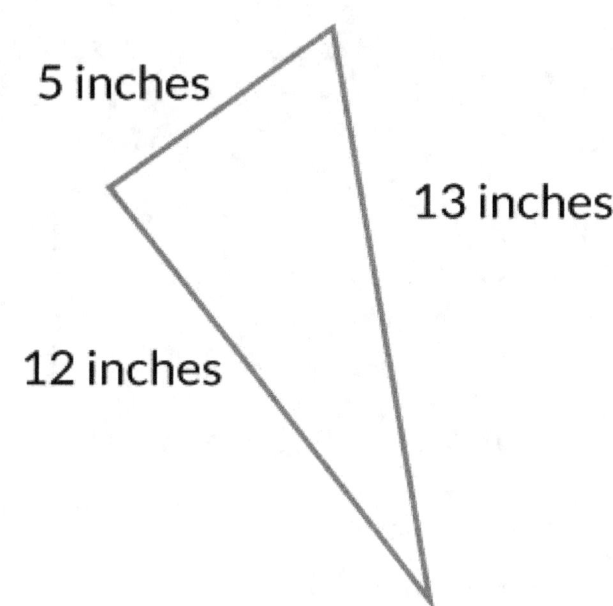

806) If the missing side is 7 inches, find the area of the following triangle.

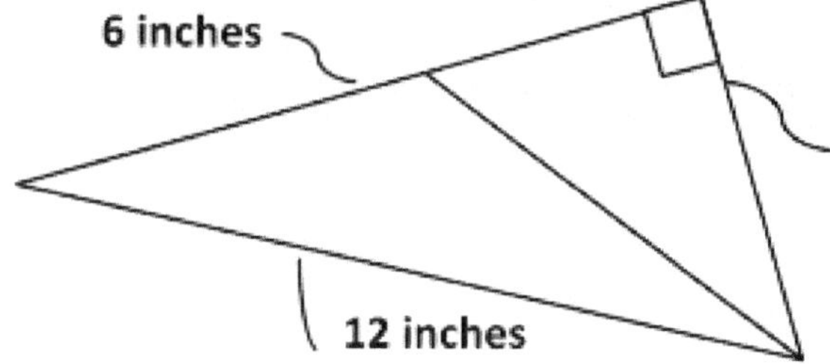

4.1 - Find the Area

807) Area of $ABC = 32\ m^2$, base $BC = 4\ m,$ the height of the triangle is_______.

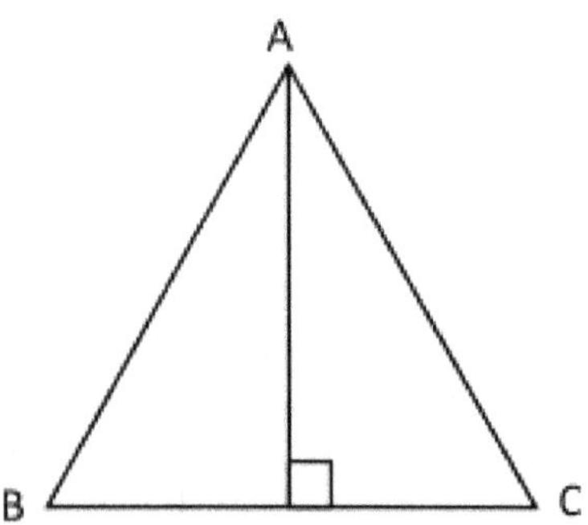

808) Determine the value of the length of the following rectangle. The area of the rectangle is $192\ cm^2$.

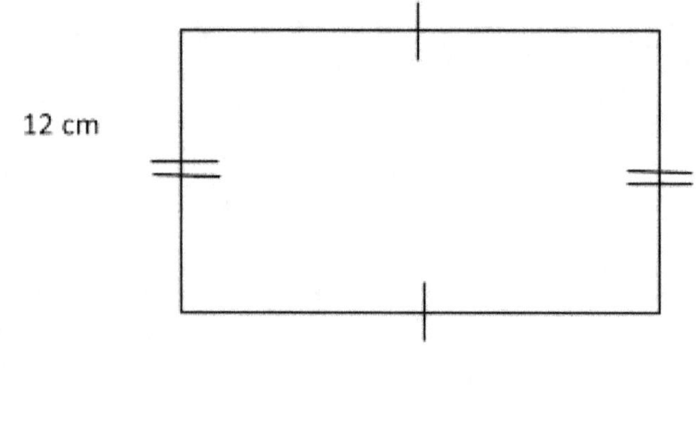

4.1 - Find the Area

809) Calculate the area of the following parallelogram.

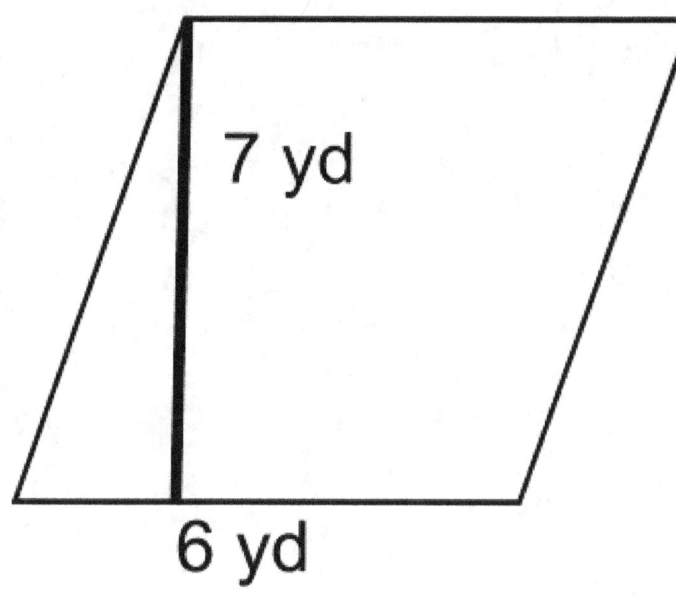

810) Calculate the area of the following square.

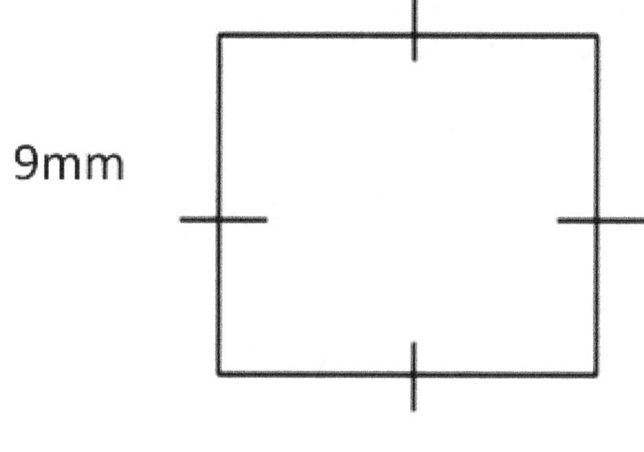

4.1 - Find the Area

811) Calculate the area of the following rhombus.

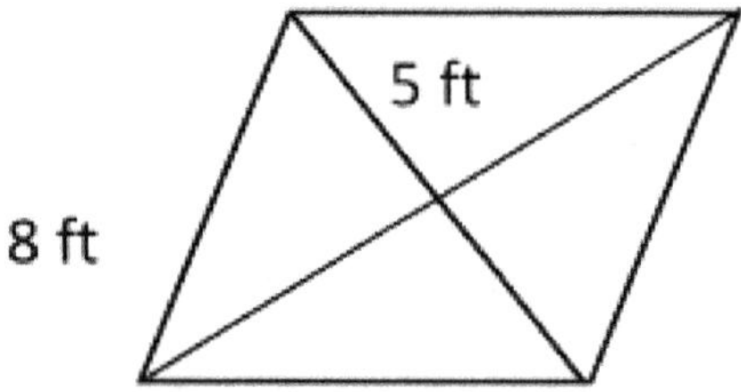

812) Calculate the area of the following rhombus.

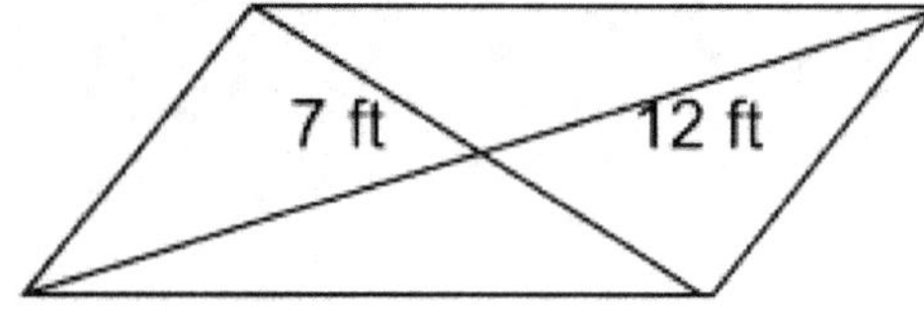

4.1 - Find the Area

813) Determine the area of the trapezoid.

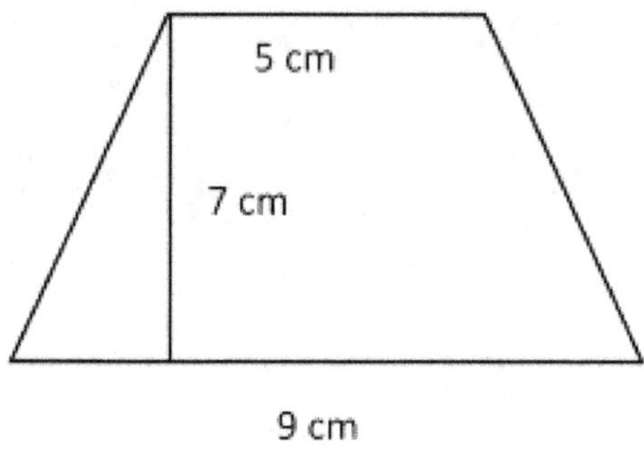

814) Calculate the area of the compound figure.

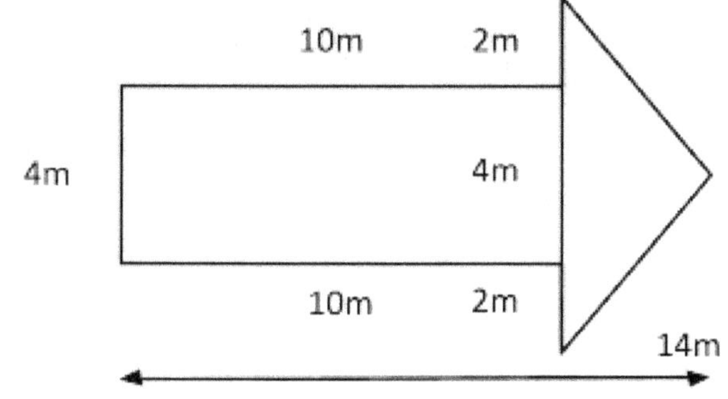

4.1 - Find the Area

815) Determine the area of the trapezoid.

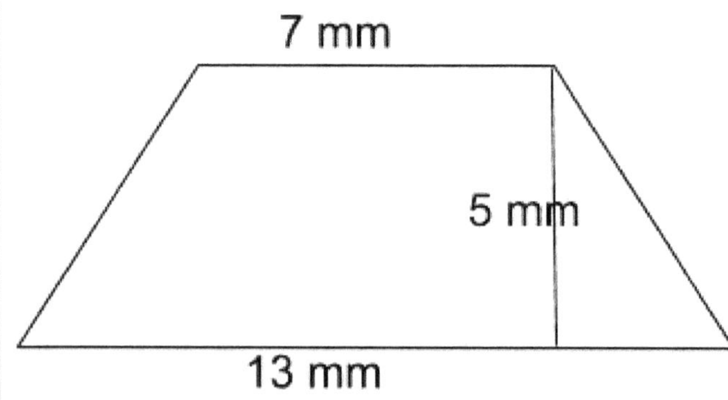

816) Determine the area of the following figure.

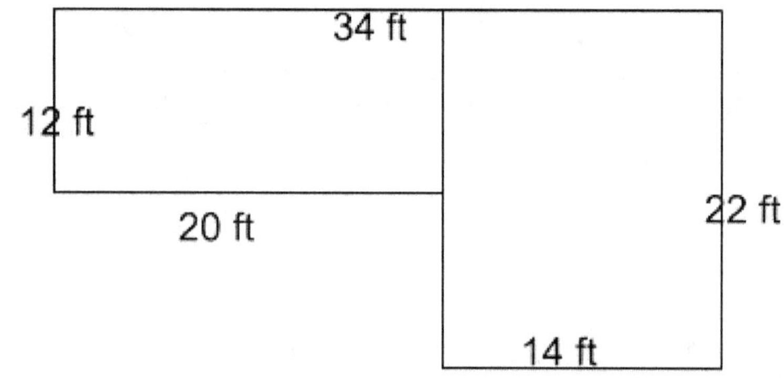

4.1 - Find the Area

817) Calculate the area of the following parallelogram.

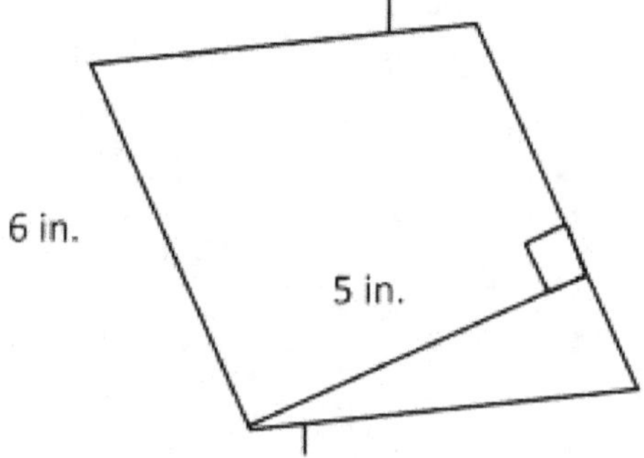

818) Calculate the area of the following parallelogram.

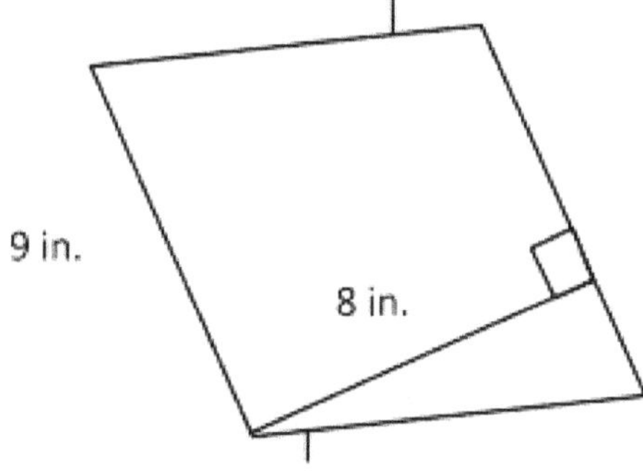

4.1 - Find the Area

819) Calculate the height of the following parallelogram. The area of the parallelogram is $45\ in^2$.

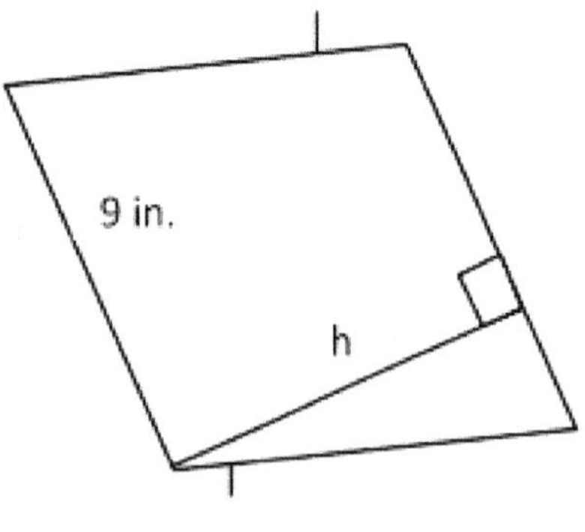

820) Find the height of the triangle. The area of the triangle is 30 m^2.

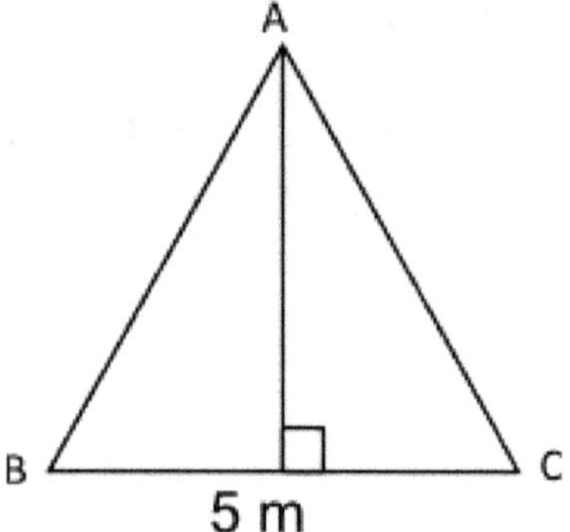

4.2 - Find the Volume

The *volume* is the amount of space measured in cubic units that an object occupies.

EXAMPLE: What is the volume of the following object?

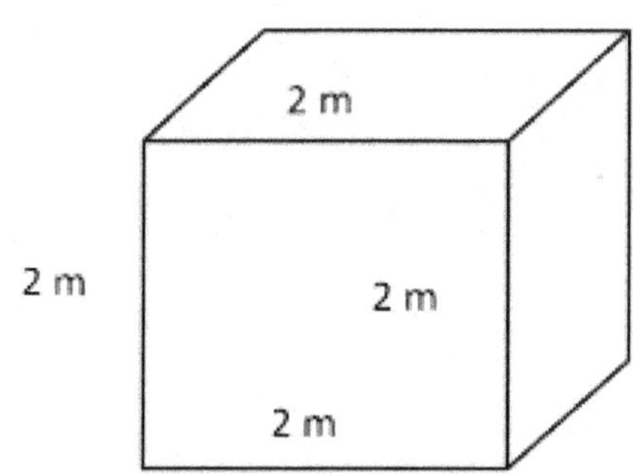

Solution: Here, all the sides of the object are of equal lengths. Therefore, it's a cube. The volume of a cube is shown as follows:

$$V = (length)^3 = (2\,m)^3 = 8\,m^3$$

Answer: **8** m^3

EXAMPLE: Find the volume of the rectangular prism.

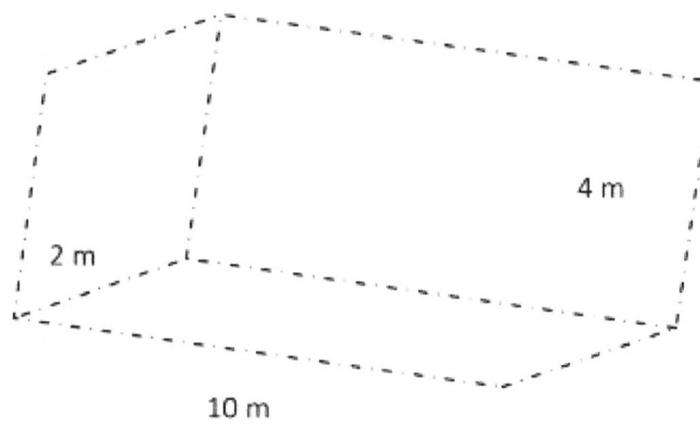

Solution: The volume of a rectangular prism is shown as follows:

V = length × width × height = 10 m × 2 m × 4 m = 80 m^3

Answer: 80 m^3

4.2 - Find the Volume

821) What is the volume of the following object?

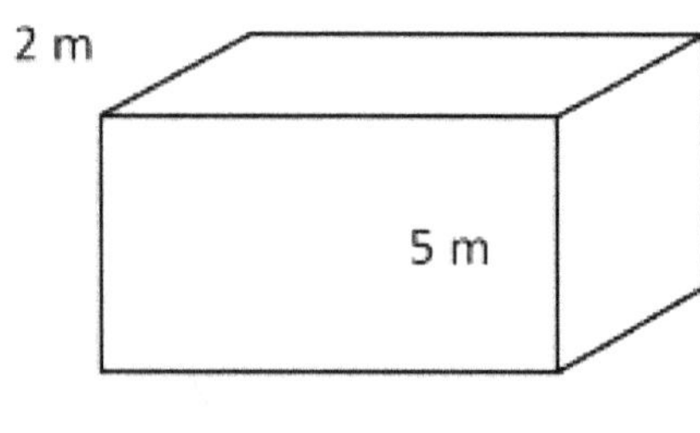

822) What is the volume of the following object?

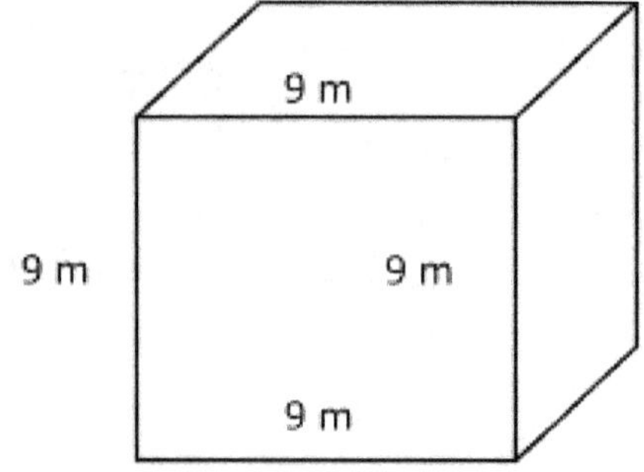

4.2 - Find the Volume

---

823) Determine the volume of the rectangular prism.

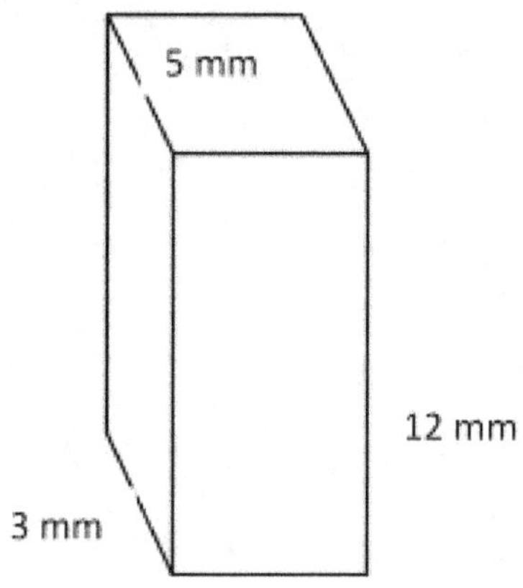

---

824) If the volume of the following object is $288\ ft^3$, then find the value of b.

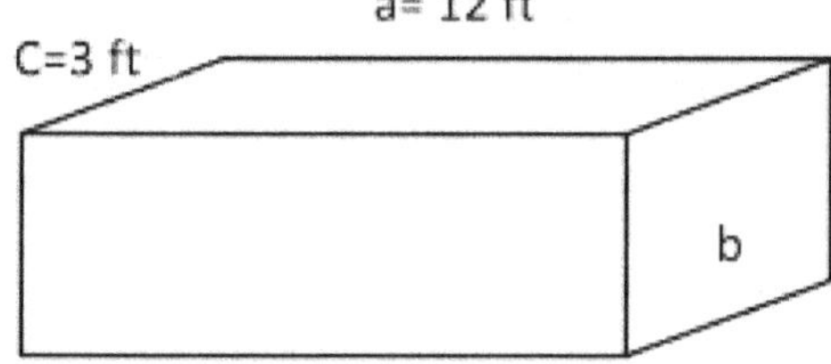

4.2 - Find the Volume

825) If the volume of a cube is $343\ cm^3$, then find the length of each side of the cube.

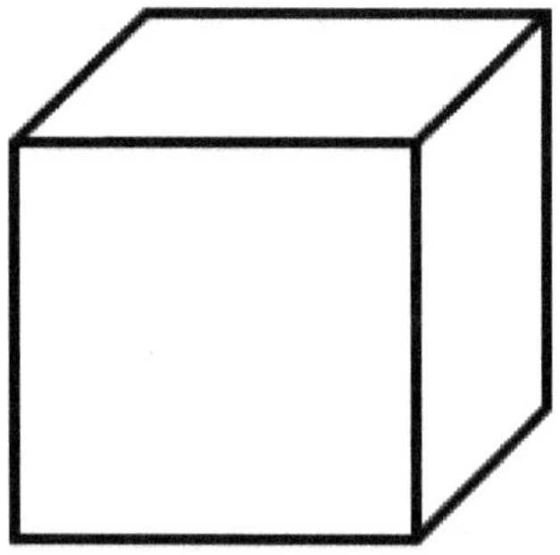

826) Liana got a gift box with a volume of $432\ yd^3$. The length and the width of the box are $9\ yd$ and $6\ yd$, respectively. Find the height of the box.

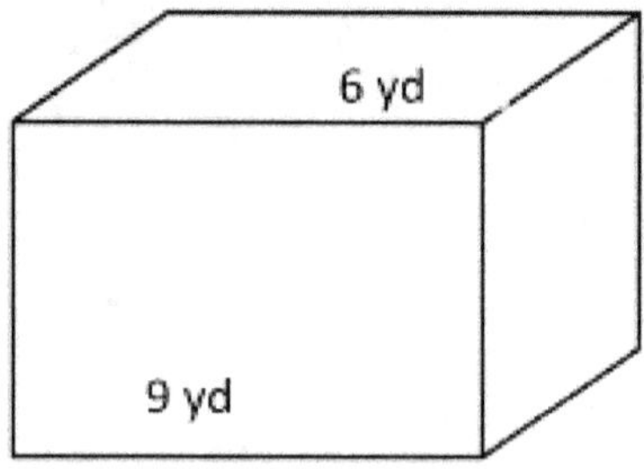

4.2 - Find the Volume

827) Zach wants to make a cubic object which is made of iron. He also wants to keep each side of the object equal. He chooses the side to be 25 mm. What would be the volume of the object?

828) Find the volume of an object, where $a = 4\ inches,\ \ b = 9\ inches,\ \ c = 11\ inches$ are the lengths of the objects on three sides.

4.2 - Find the Volume

829) What is the volume of the object below?

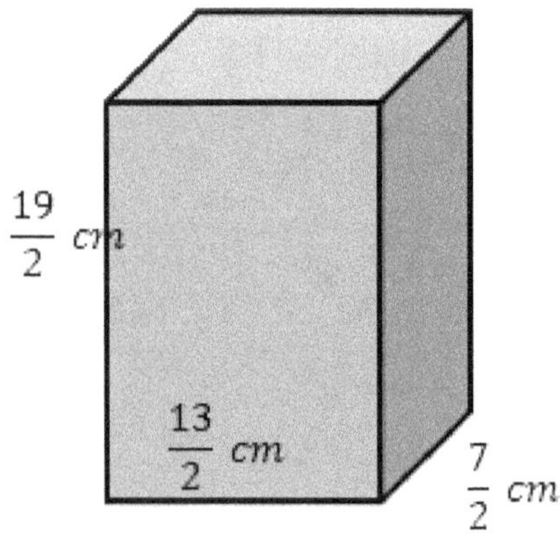

830) What is the volume of the object below?

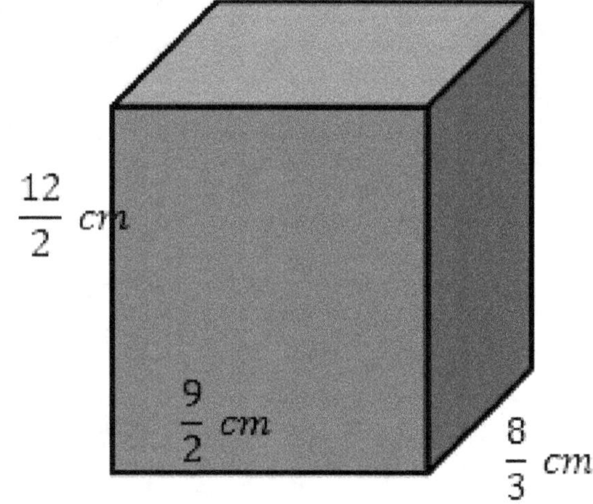

4.2 - Find the Volume

831) Find the volume of the cylinder using $V = \pi r^2 h$.

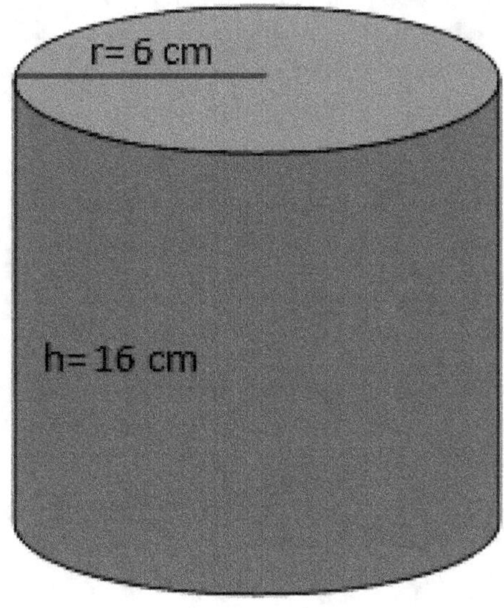

832) Find the volume of the following pipe. (Note: radius = 1/2 m)

4.2 - Find the Volume

833) Find the volume of the rectangular prism.

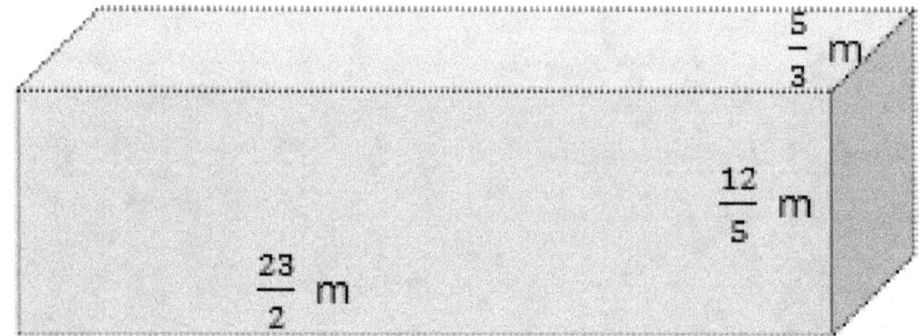

834) Find the volume of the rectangular prism.

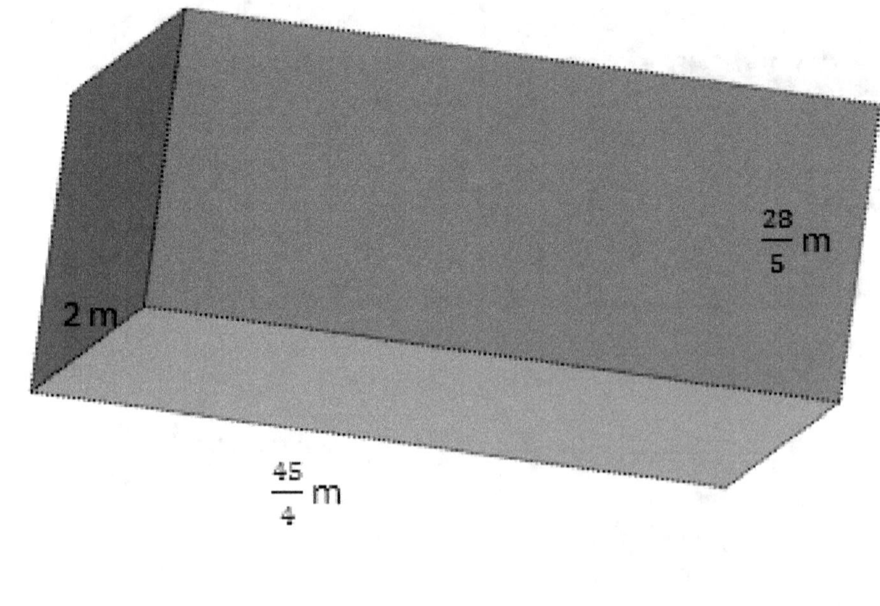

4.2 - Find the Volume

835) Zachary wants to make a cubic object which is made of iron. Also, he wants to keep each side of the object equal. He chooses the side to be 6 mm. How much iron does he need to make this cubic object?

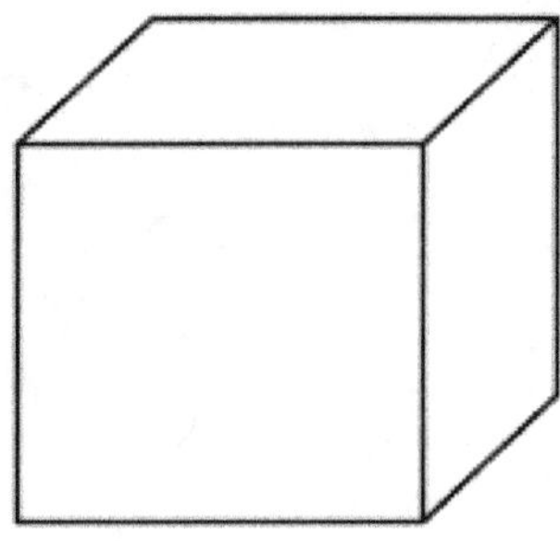

836) Jane plans to make a big cubic object made of 50 small cubic blocks. If the volume of the big one is $4,870\ cm^3$, what is the volume of each small block?

4.2 - Find the Volume

837) A rectangular prism-shaped structure of $5\ m^3$ volume costs \$23. If Sara wants to buy a structure of similar shape but of $120\ m^3$ volume, then how much money would she have to spend?

838) What is the volume of the following rectangular prism?

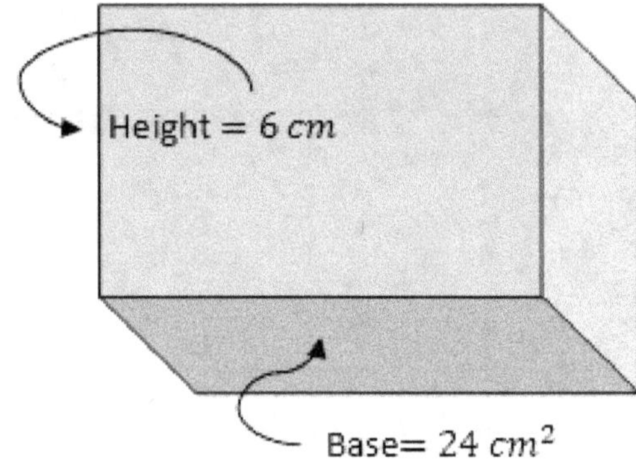

4.2 - Find the Volume

---

839) What is the volume of the following rectangular prism?

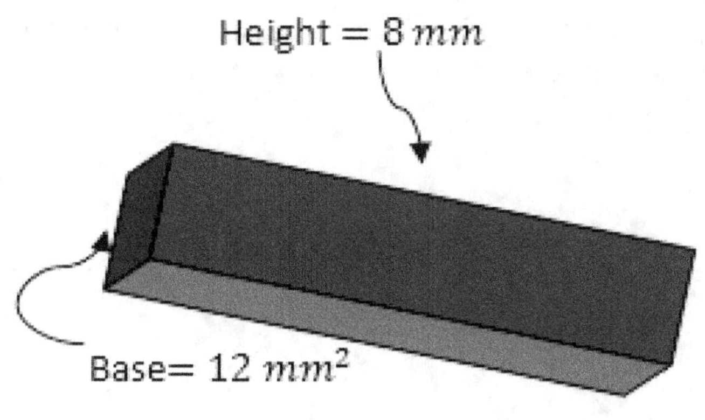

---

840) The following is a rectangular prism-shaped box. If Rosie fills this box with small cubic boxes with a side length of $0.5\ m,$ how many boxes would she need?
Dimensions = 2 m by 76/3 m by 57/4 m.

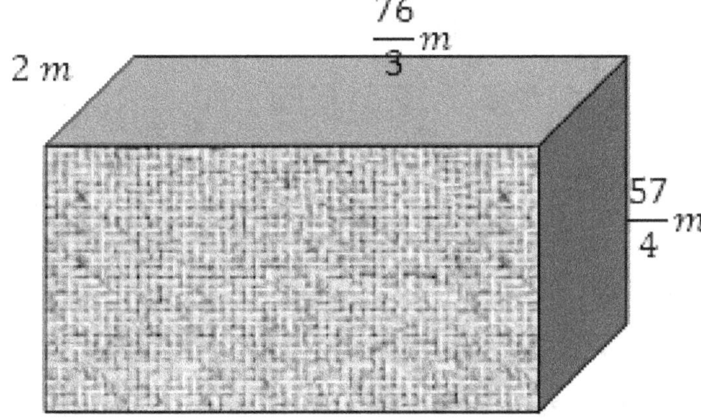

4.3 - Coordinate Plane

A *coordinate plane* is a two-dimensional plane formed by the intersection of two lines. The horizontal line is known as the *x-axis*, while the vertical line is known as the *y-axis*.

EXAMPLE: Write the coordinates of the following triangle.
Solution: To find a coordinate pair, first find the x-coordinate by beginning at the point and following a vertical line either up or down to the x-axis. Then, find the y-coordinate by following a horizontal line from the point either left or right to the y-axis.

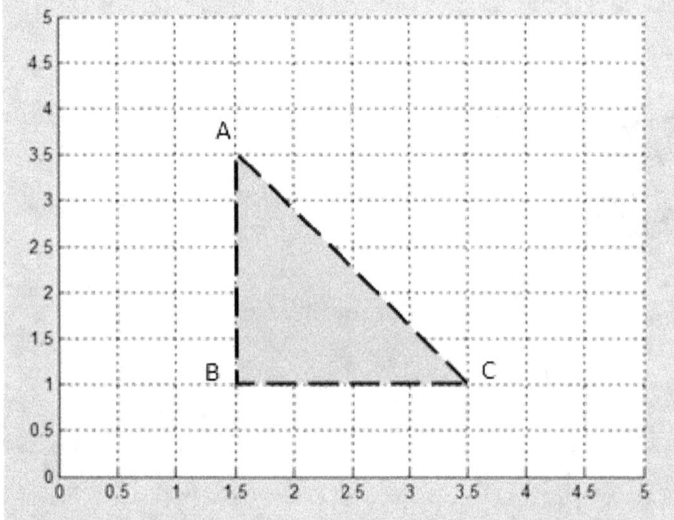

Answer:

$A \equiv (1.5, 3.5), B \equiv (1.5, 1), C \equiv (3.5, 1)$

EXAMPLE: From the graph below, determine whether x > y or y > x.

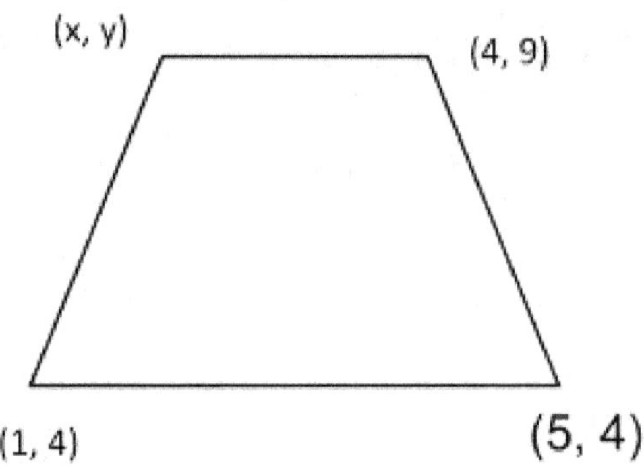

Solution: Here, the given shape is a trapezoid. There are two parallel sides. Therefore, *y* must be equal to 9, and *x* will take a value that is greater than 1 but less than 4. Hence, 1 < x < 4, and y = 9.

Answer: **y > x**

4.3 - Coordinate Plane

---

841) To draw a right triangle, what coordinates may be the third coordinate for $y = 6$?

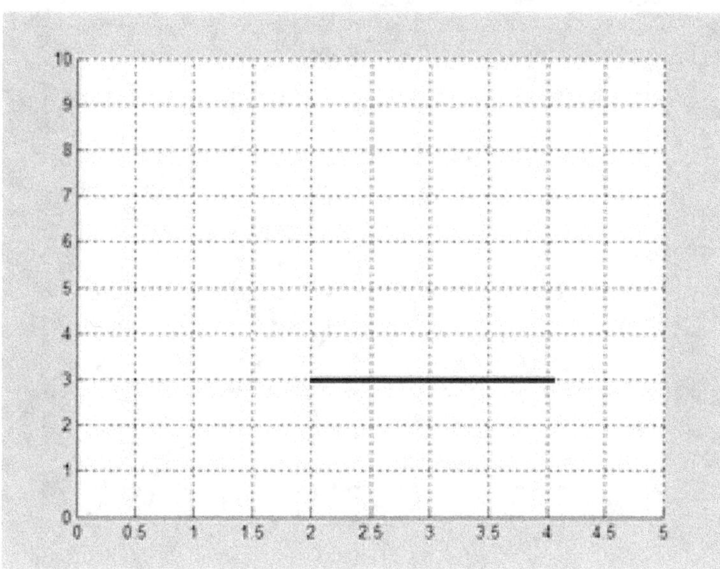

---

842) Write the coordinate that is required to form the following rectangle.

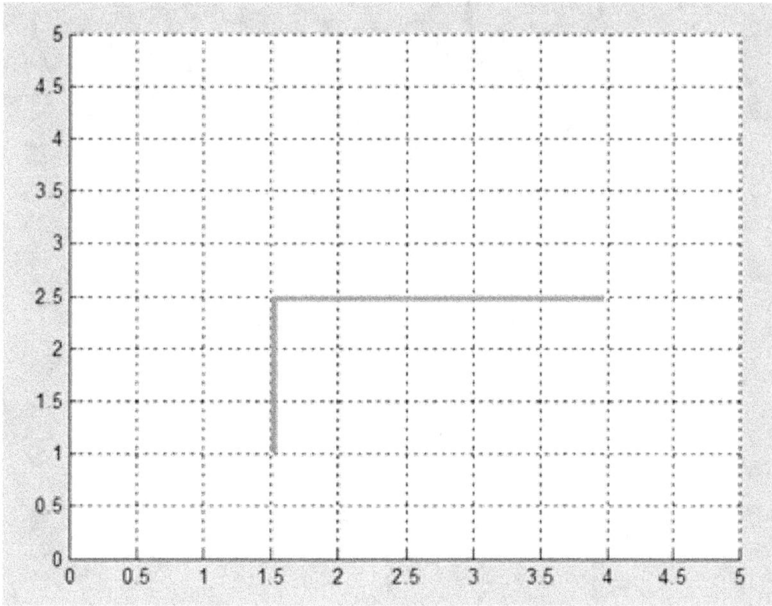

4.3 - Coordinate Plane

843) Find the area of the following triangle.

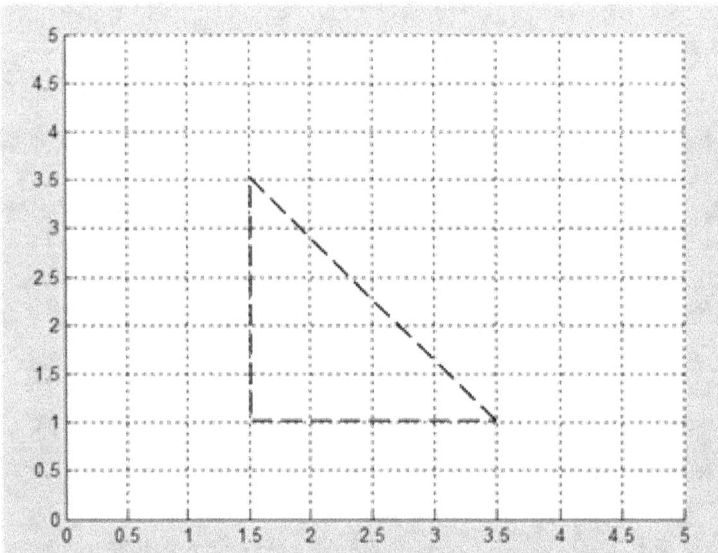

844) Write the coordinates of the following triangle.

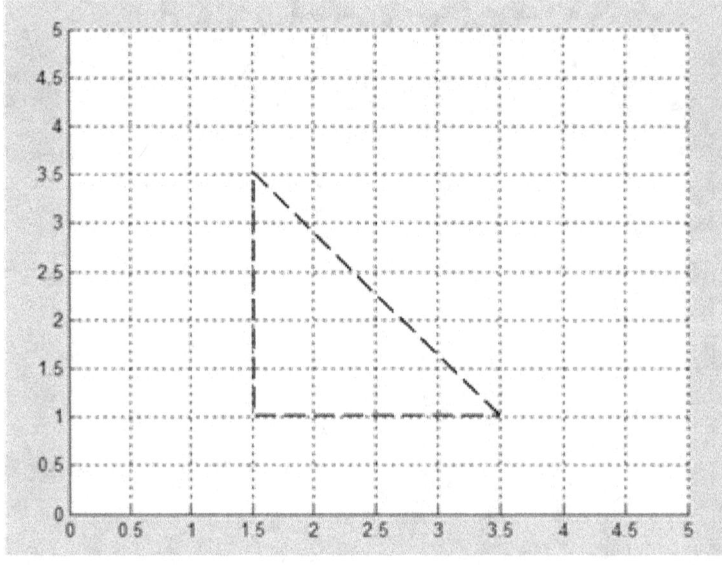

4.3 - Coordinate Plane

845) Write the coordinates of the following trapezoid.

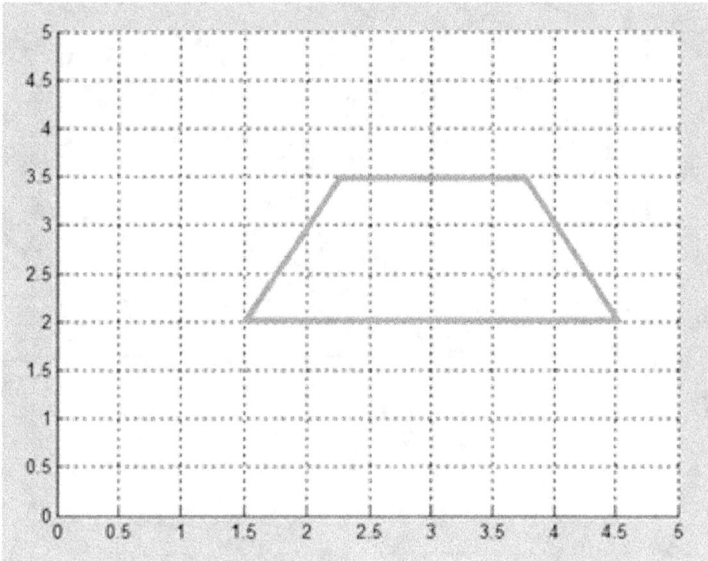

846) Draw a rectangle placed on the coordinates (1, 1), (4, 1), (1, 3), and (4, 3). What is the area of the rectangle?

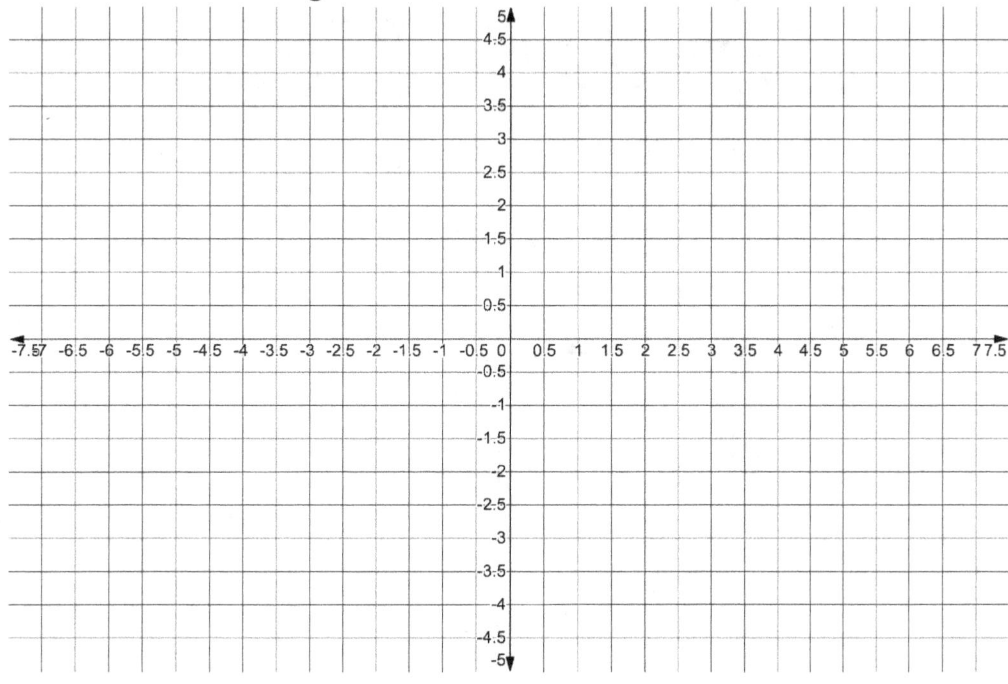

4.3 - Coordinate Plane

847) Draw a rectangle placed on the coordinates (0.5, 1), (3.5, 1), (3.5, 4), and (0.5, 4). What is the area of the rectangle?

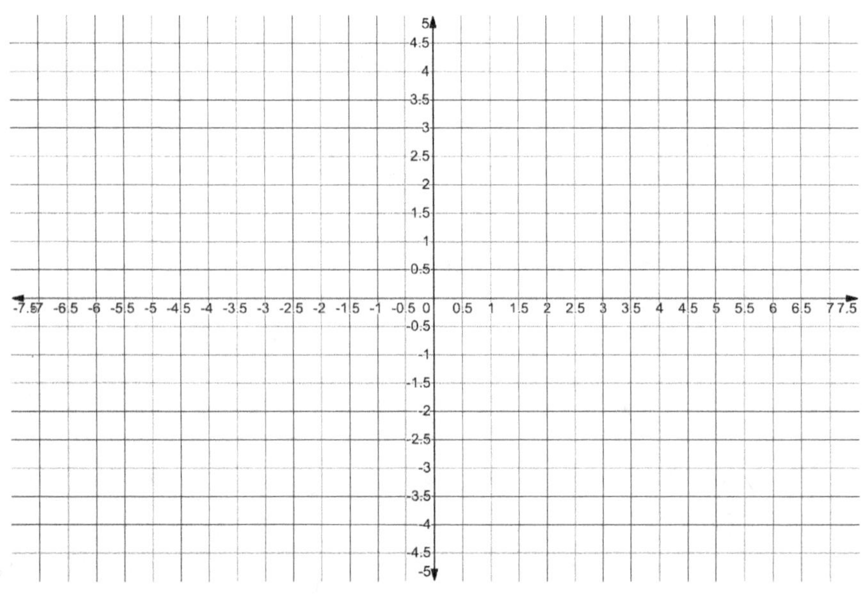

848) Draw a triangle if the given vertices are (-3, -3), (3, -3), and (0, 3).

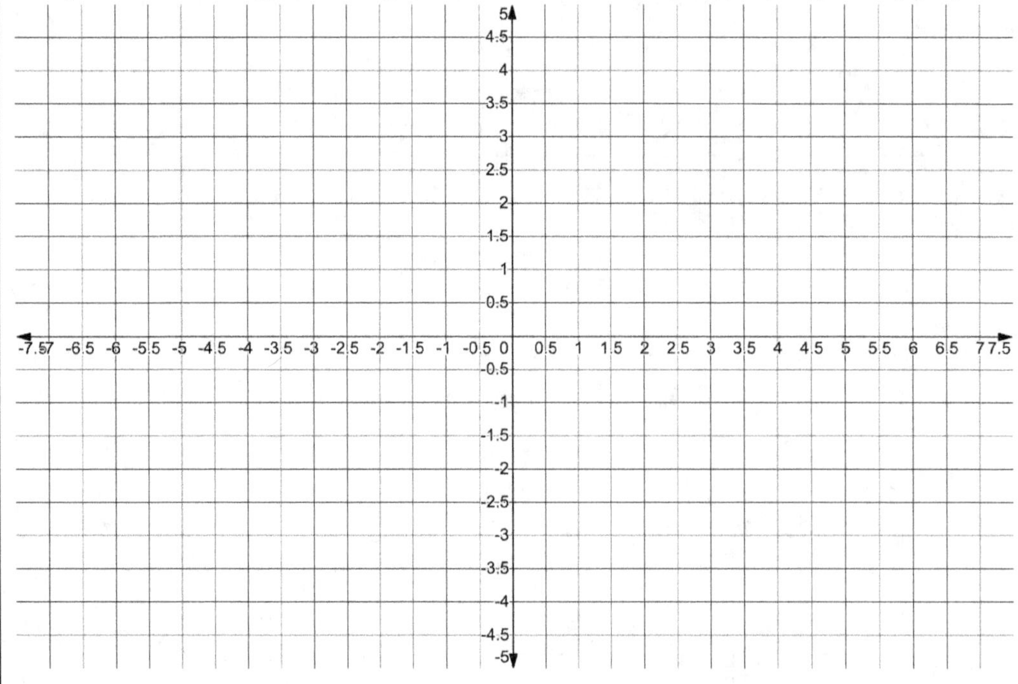

4.3 - Coordinate Plane

849) Draw a parallelogram, if the given vertices are (-3, -2), (2, -2), (-2, 2), and (3, 2). Find the area of the parallelogram.

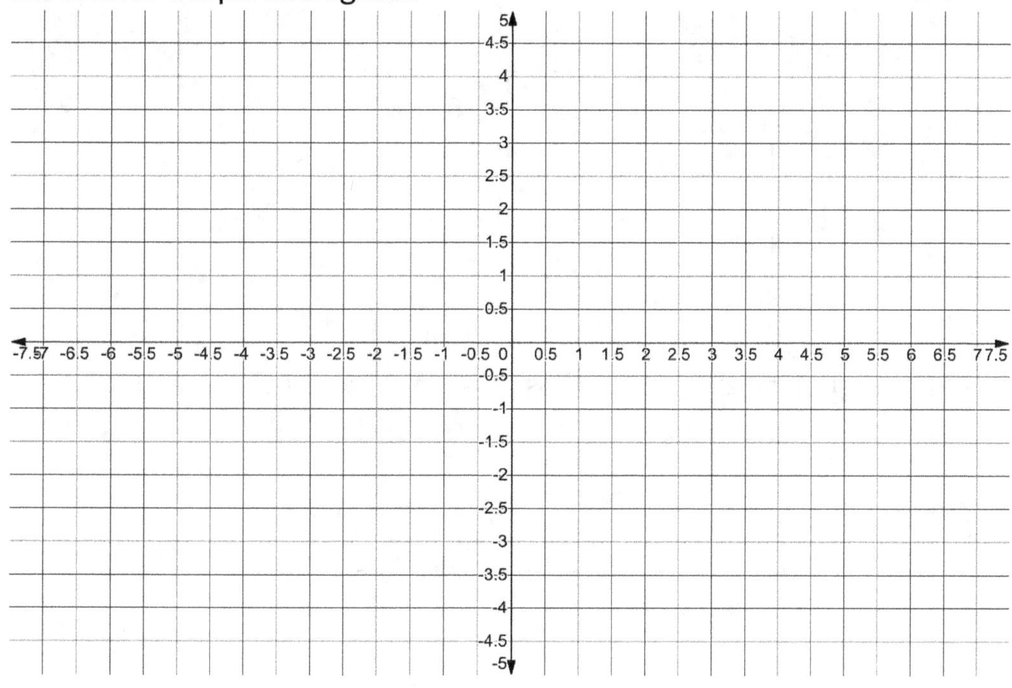

850) From the graph below, determine whether x > y or y > x.

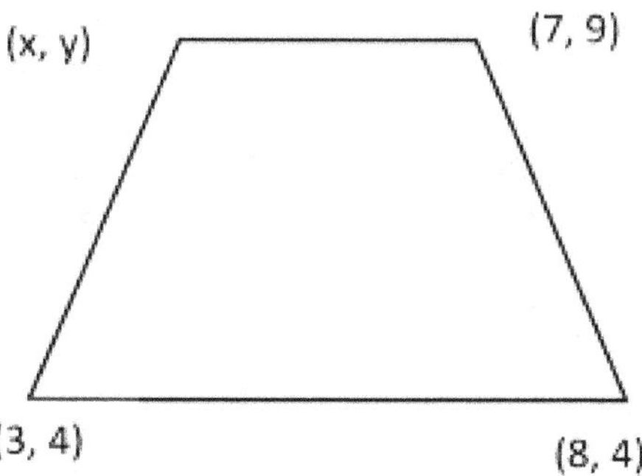

4.4 - Surface Area

The *surface area* is the sum of the areas of all faces of a three-dimensional object.

EXAMPLE: Calculate the surface area of the following cube.

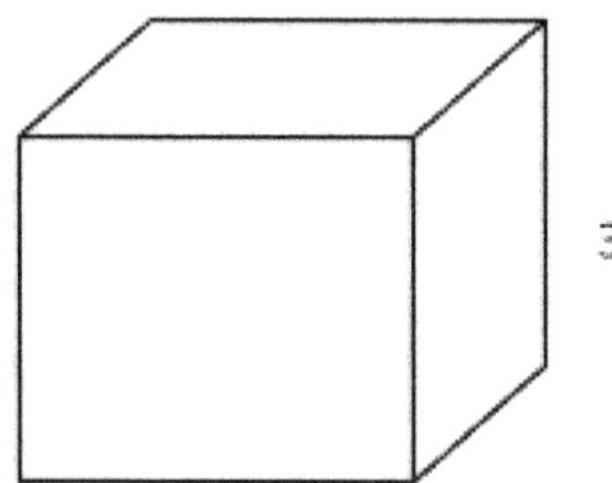

Solution: The cube has six equal faces, and each has an area of $A = s^2$. Hence, the surface area is $SA = 6 \times (3^2) = 6 \times 9 = 54$.

Answer: **54 square units**

EXAMPLE: How much paper is needed to wrap the box?

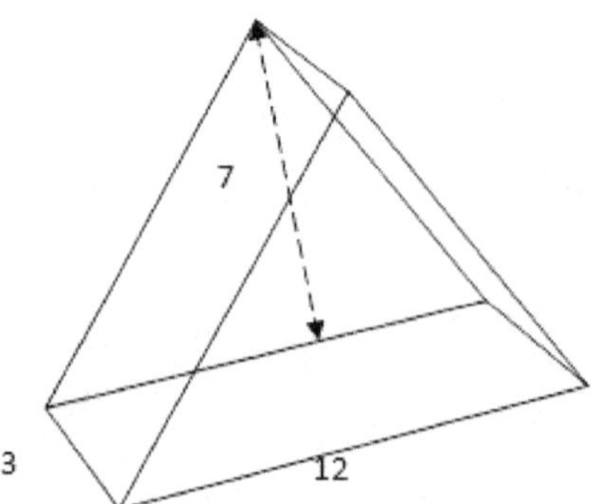

Solution: To wrap the entire box, the paper equals the surface area is required. The surface area of the triangular prism is given as follows (note that there are two equal triangles and three equal rectangles):

$$SA = 2 \times (\text{area of a triangle}) + 3 \times (\text{area of a rectangle})$$
$$SA = 2 \times (\tfrac{1}{2} \times \text{base} \times \text{height}) + 3 \times (\text{length} \times \text{width})$$
$$SA = 2 \times (\tfrac{1}{2} \times 12 \times 7) + 3 \times (12 \times 3)$$
$$SA = (2 \times 42) + (3 \times 36)$$
$$SA = 84 + 108$$
$$SA = 192$$

Answer: **192 square units** of paper is required to wrap the box.

4.4 - Surface Area

851) What is the surface area of the following cube?

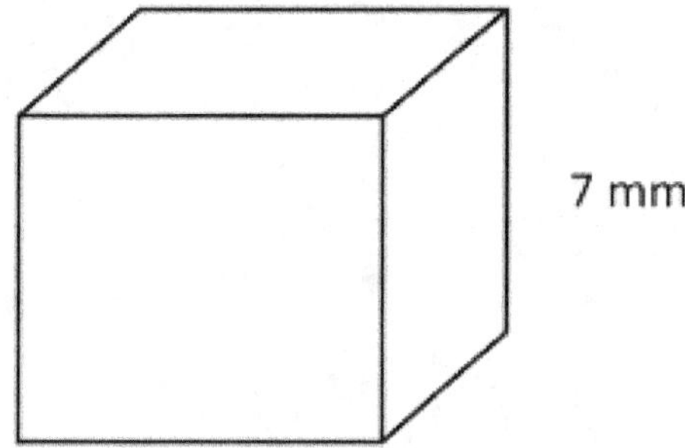

852) What is the surface area of the following rectangular prism?

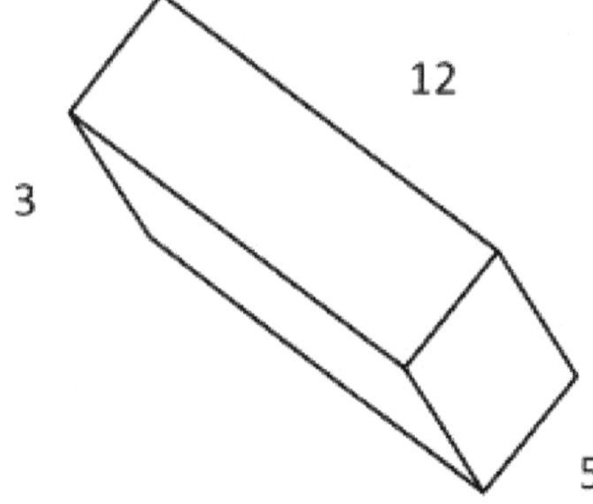

4.4 - Surface Area

853) Bill is wrapping rectangular prism-shaped gift boxes with multicolor printed wrapping paper. The length, width, and height of the boxes are 2.5 m, 0.5 m, and 1 meter. How much paper does she need to wrap each box?

854) Find the surface area of a rectangular prism that is 7 cm long, 2 cm wide, and the height of which is 5.6 cm.

4.4 - Surface Area

---

855) Calculate the surface area of the following cube.

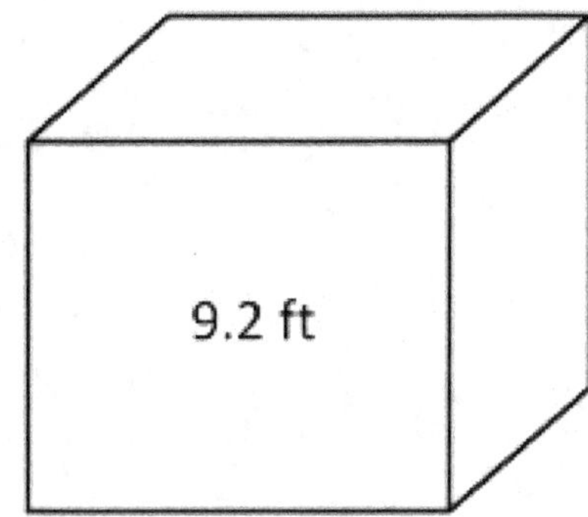

---

856)What shape can be formed from the figure below? What will be the shape's surface area?

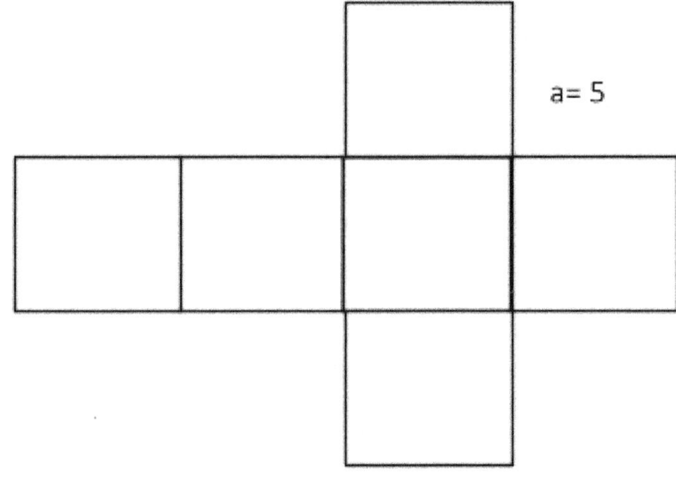

4.4 - Surface Area

857) What shape can be formed from the figure below?

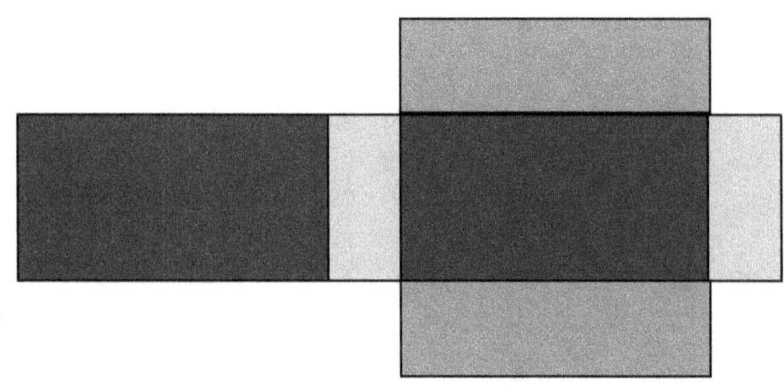

858) Draw a net (unfolded version) that can construct the shape below.

4.4 - Surface Area

859) What shape can be formed from the following net?

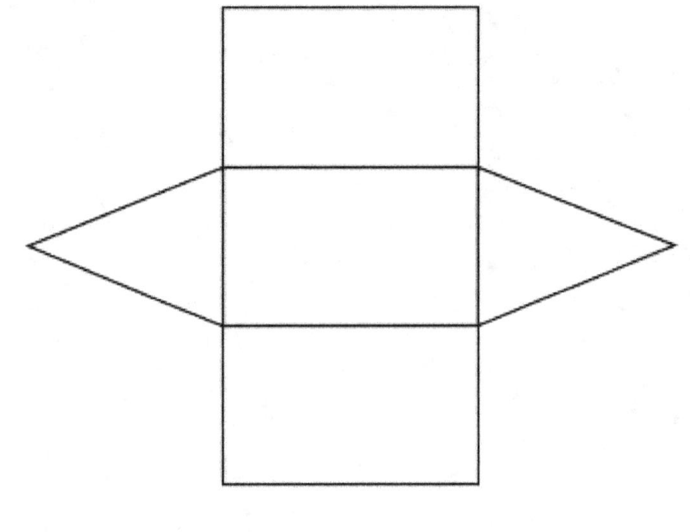

860) Use the following net to calculate the surface area of the rectangular prism.

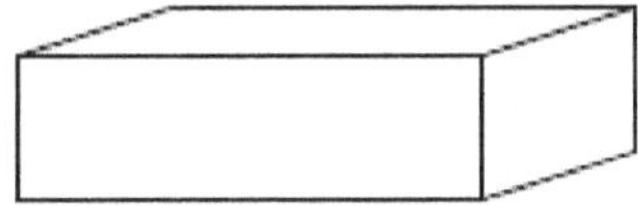

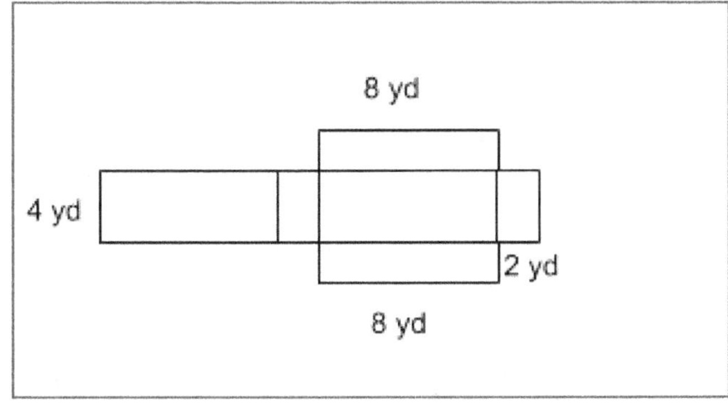

4.4 - Surface Area

861) What shape can be formed from the following net?

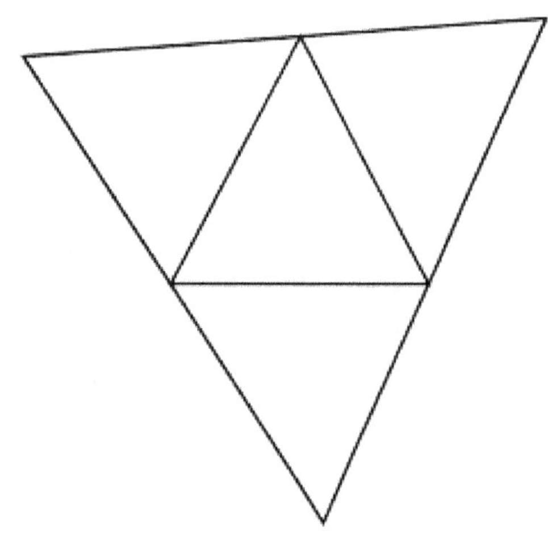

862) Draw a net that can construct the shape below.

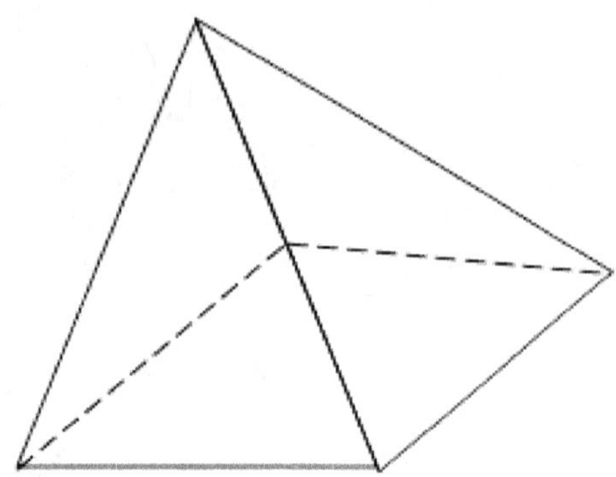

4.4 - Surface Area

863) Holly is wrapping a chocolate box of the following shape. How much paper does she need to wrap the box?

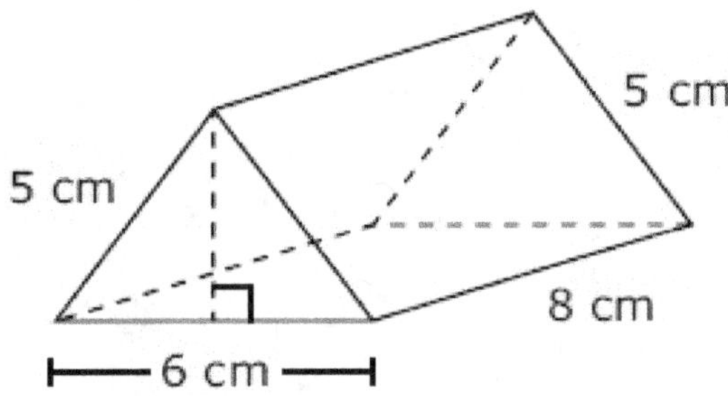

864) Joana went to a jewelry shop to place an order of a pair of earrings. She wanted it to be of the following square shape. If it's made of silver, how much silver did she need for each of the earrings?

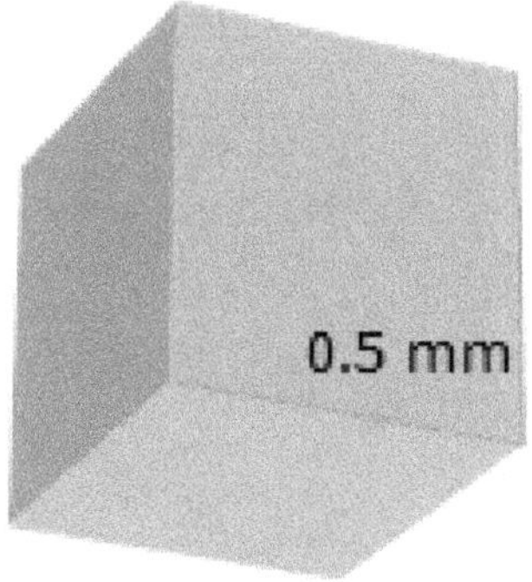

4.4 - Surface Area

---

865) To wrap a clay art of the following shape, how much paper is needed?

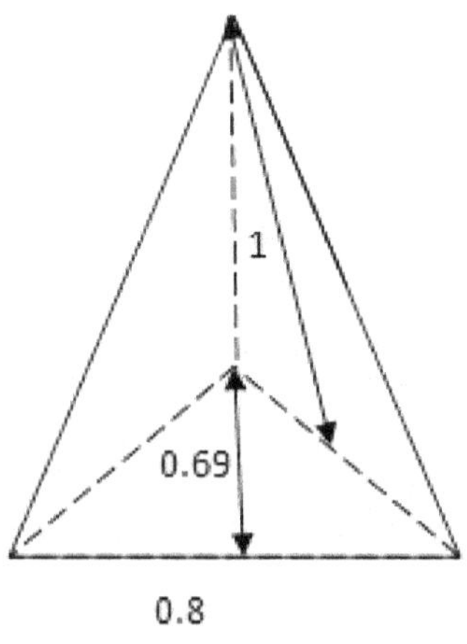

---

866) Write an expression to calculate the surface area by adding face areas.

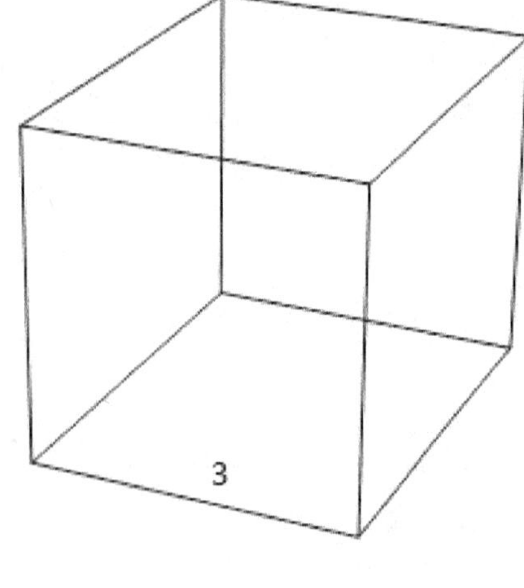

4.4 - Surface Area

---

867) Write an expression to calculate the surface area by adding face areas.

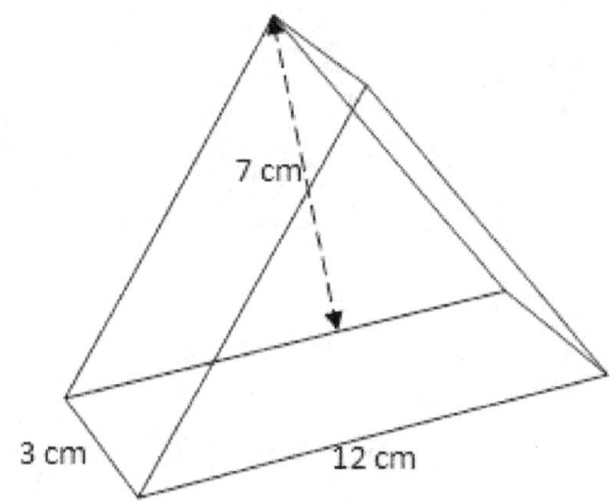

---

868) Write an expression to calculate the surface area by adding face areas.

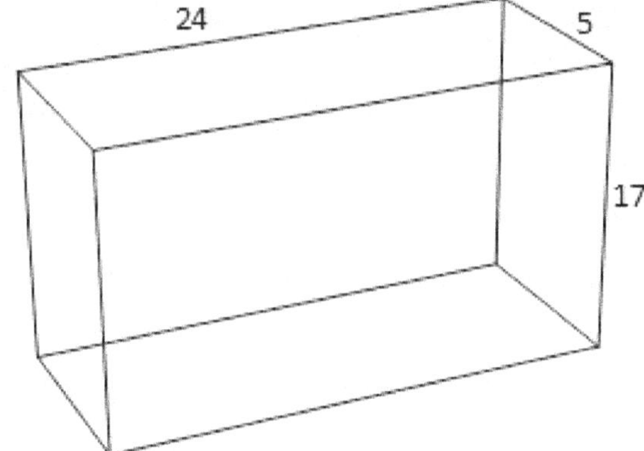

4.4 - Surface Area

869) Write an expression to calculate the surface area by adding face areas.

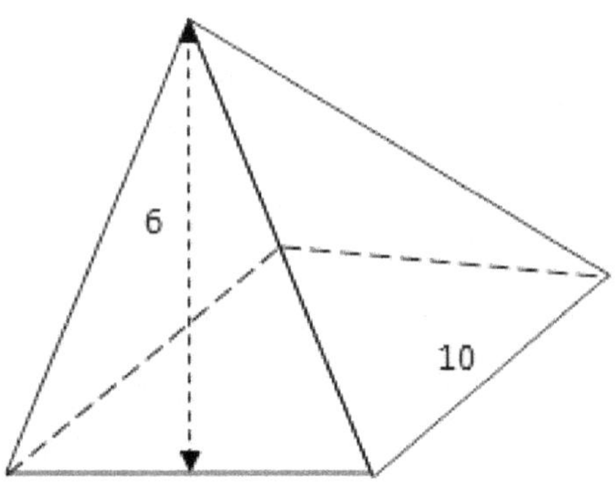

870) Write an expression to calculate the surface area by adding face areas.

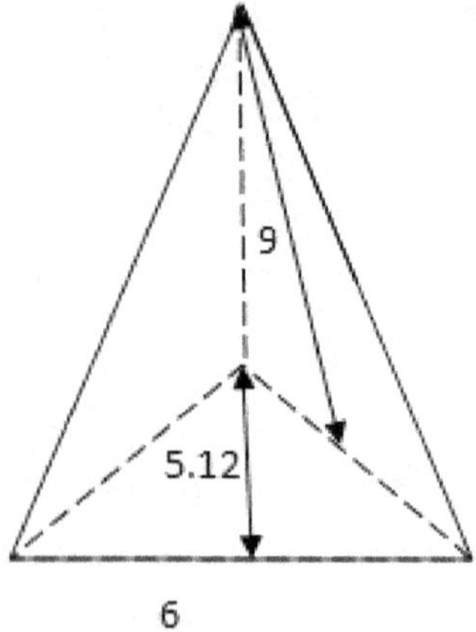

5.1 - Identify Statistical Questions

A *statistical question* refers to a question that can be answered by gathering a variety of data. For example, "What time did I get up this morning?" is a non-example of a statistical question. This will turn as statistical if the question is, "What time did the students in this class get up in the morning?".

EXAMPLE: Determine whether the question is statistical or not: *How tall are you?*

Solution: Remember that a statistical question must collect data that vary. Here, the answer is just your height; hence, there is no variability in the data. Thus, it is not a statistical question.

Answer: **No**

EXAMPLE: Determine whether the question is statistical or not: *How tall in inches, Jacob on his last birthday?*

Solution: Remember that a statistical question must collect data that vary. Here, the answer is just Jacob's height; hence, there is no variability in the data. Thus, it is not a statistical question.

Answer: **No**

EXAMPLE: Determine whether the question is statistical or not: *How tall are you among your class?*

Solution: To answer this question, you need to know your height and your classmates' heights. There is variability in the data. Thus, it is a statistical question.

Answer: **Yes**

5.1 - Identify Statistical Questions

871) Determine whether the question is statistical or not: *How many months are left this year?*	872) Determine whether the question is statistical or not: *What classes do you take this semester?*
873)Determine whether the question is statistical or not: *How old are you?*	874) Determine whether the question is statistical or not: *Sally is the tallest person among his friends.*
875)Determine whether the question is statistical or not: *How far do you drive this morning?*	876) Determine whether the question is statistical or not: *How many umbrellas do they have?*
877) Determine whether the question is statistical or not: *What is the average age of three students?*	878) Determine whether the question is statistical or not: *How many hours did you work yesterday?*
879)Determine whether the question is statistical or not: *What is the area of a rectangle?*	880)Determine whether the question is statistical or not: *Who is the oldest student in this class?*

5.2 - Distribution

A *distribution* in statistics shows the possible values for a certain variable and how often these values occur.

EXAMPLE: Find the median of the data in the box plot below

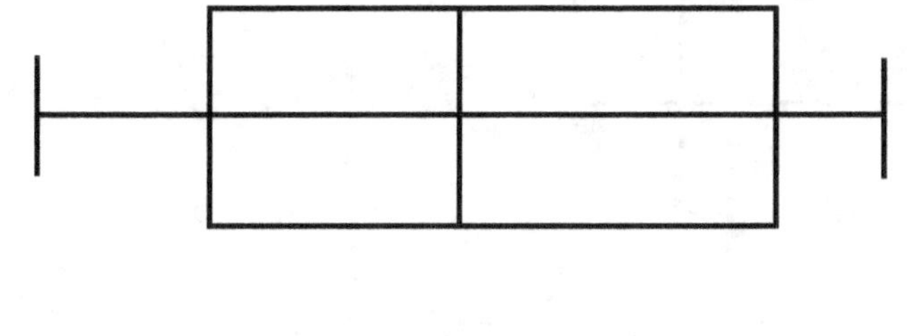

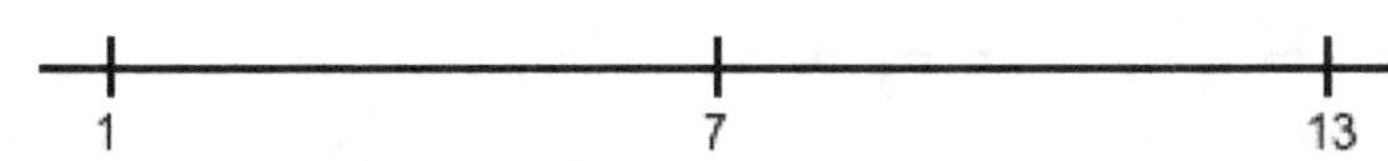

Solution: The *median* is the middle value of a box plot. Thus, the median of the data set is 7.

Answer: **7**

EXAMPLE: What is the shape of the data in the box plot below. Is it symmetrical or not?

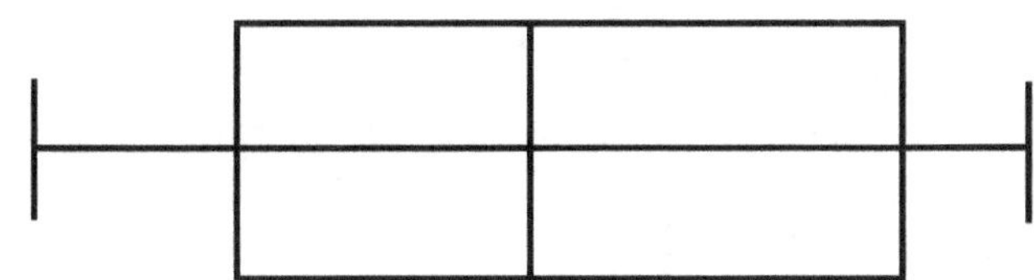

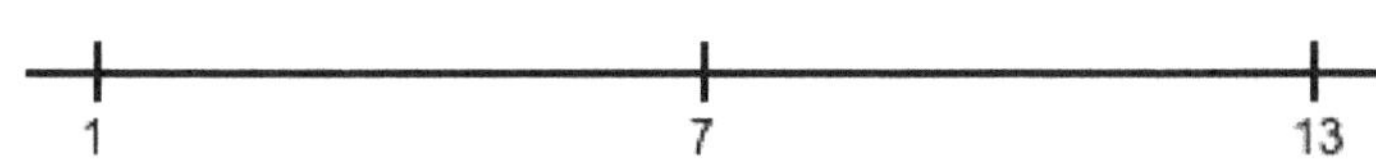

Answer: Notice how the "tail" on the left is longer than the right. Therefore, the data is *not symmetrical*.

5.2 - Distribution

881) Find the maximum value of the following box plot.

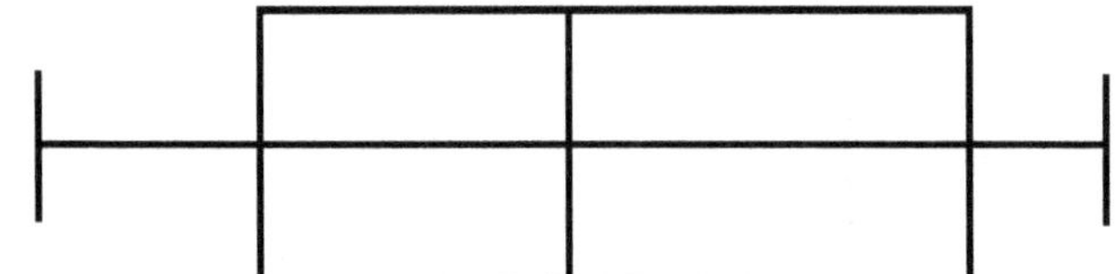

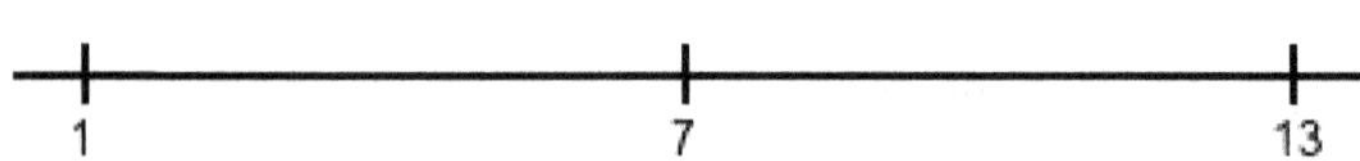

882) Find the median of the data in the box plot below.

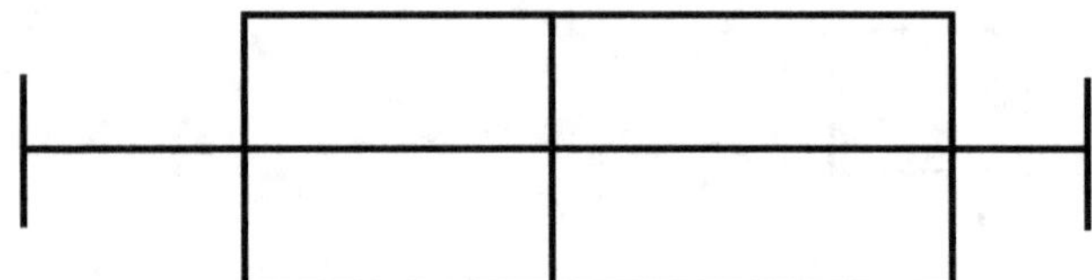

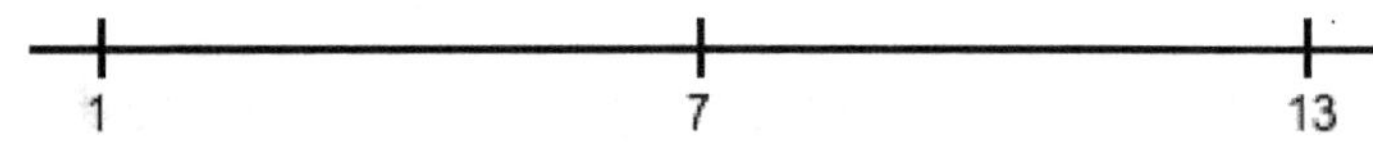

5.2 - Distribution

883) Find the lower quartile of the data in the box plot below.

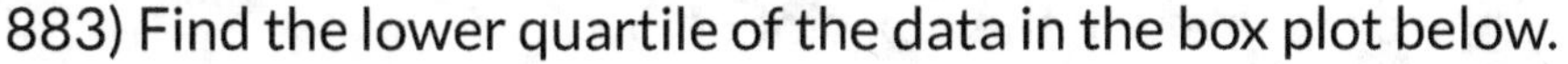
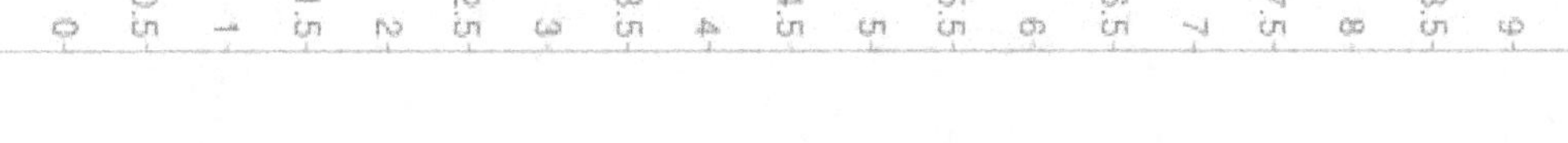
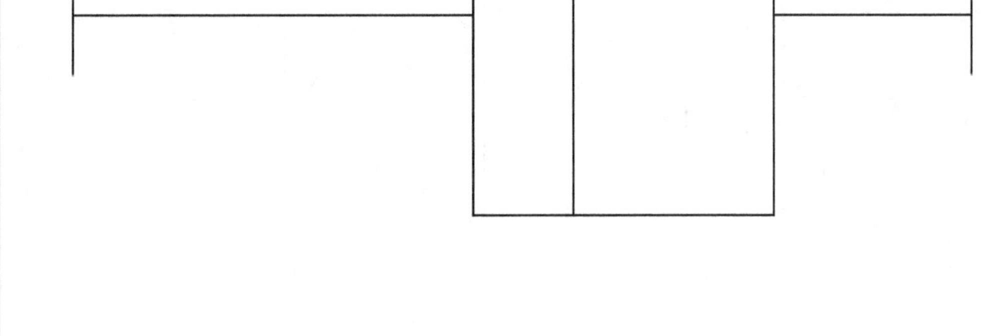

884) Find the upper quartile of the data in the box plot below.

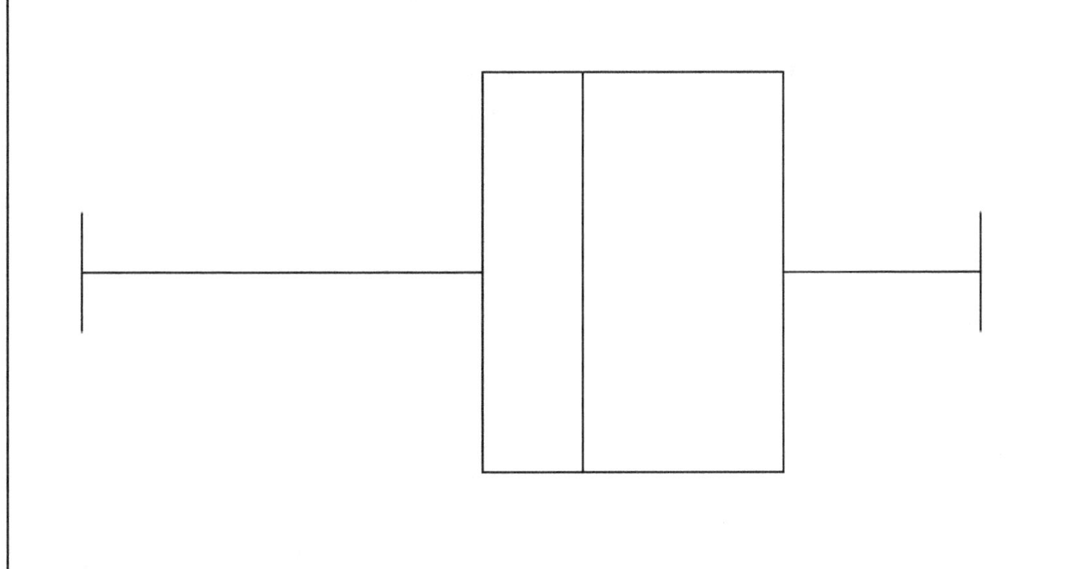

5.2 - Distribution

885) Find the interquartile range of the data in the box plot below.

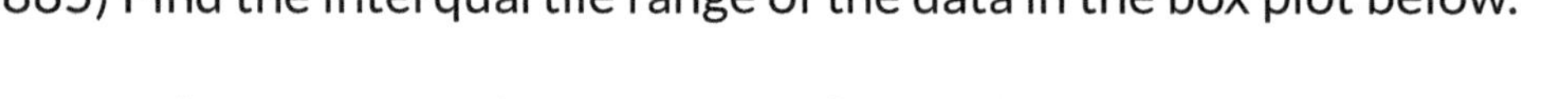

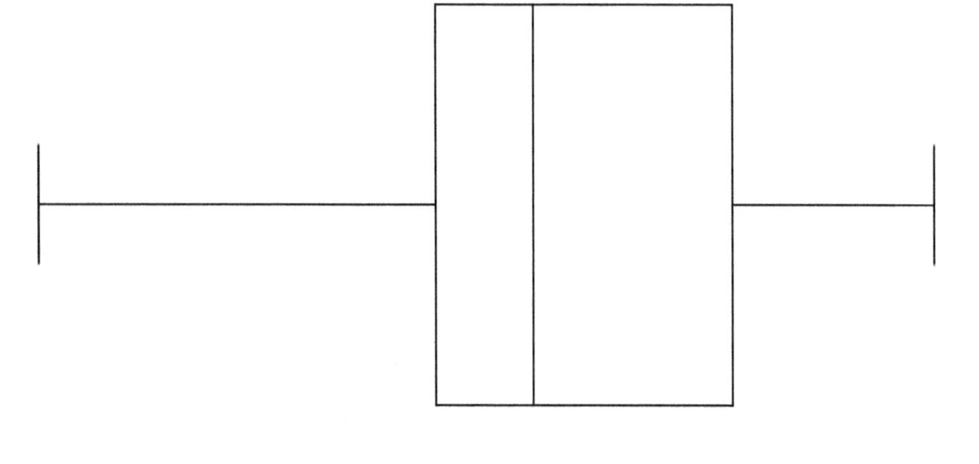

886) Find the range of the data in the box plot below.

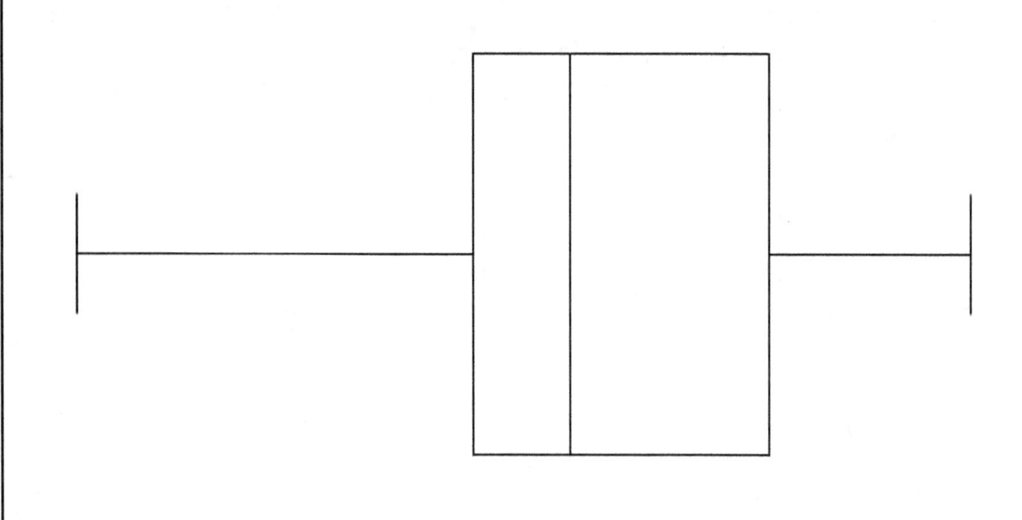

5.2 - Distribution

887) Find the median of the data in the box plot below.

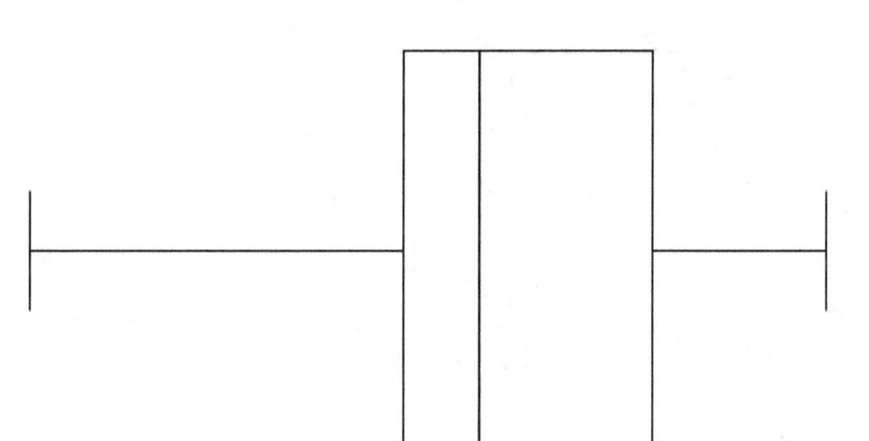

888) Find the range of the data in the box plot below.

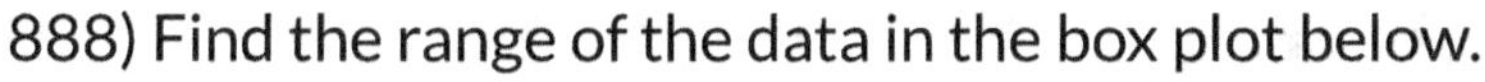

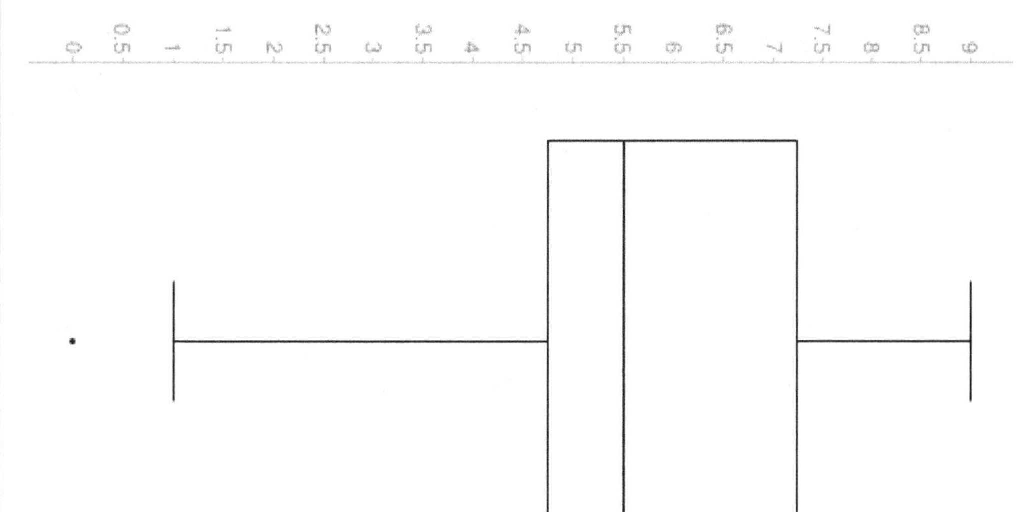

5.2 - Distribution

889) What is the shape of the following distribution?

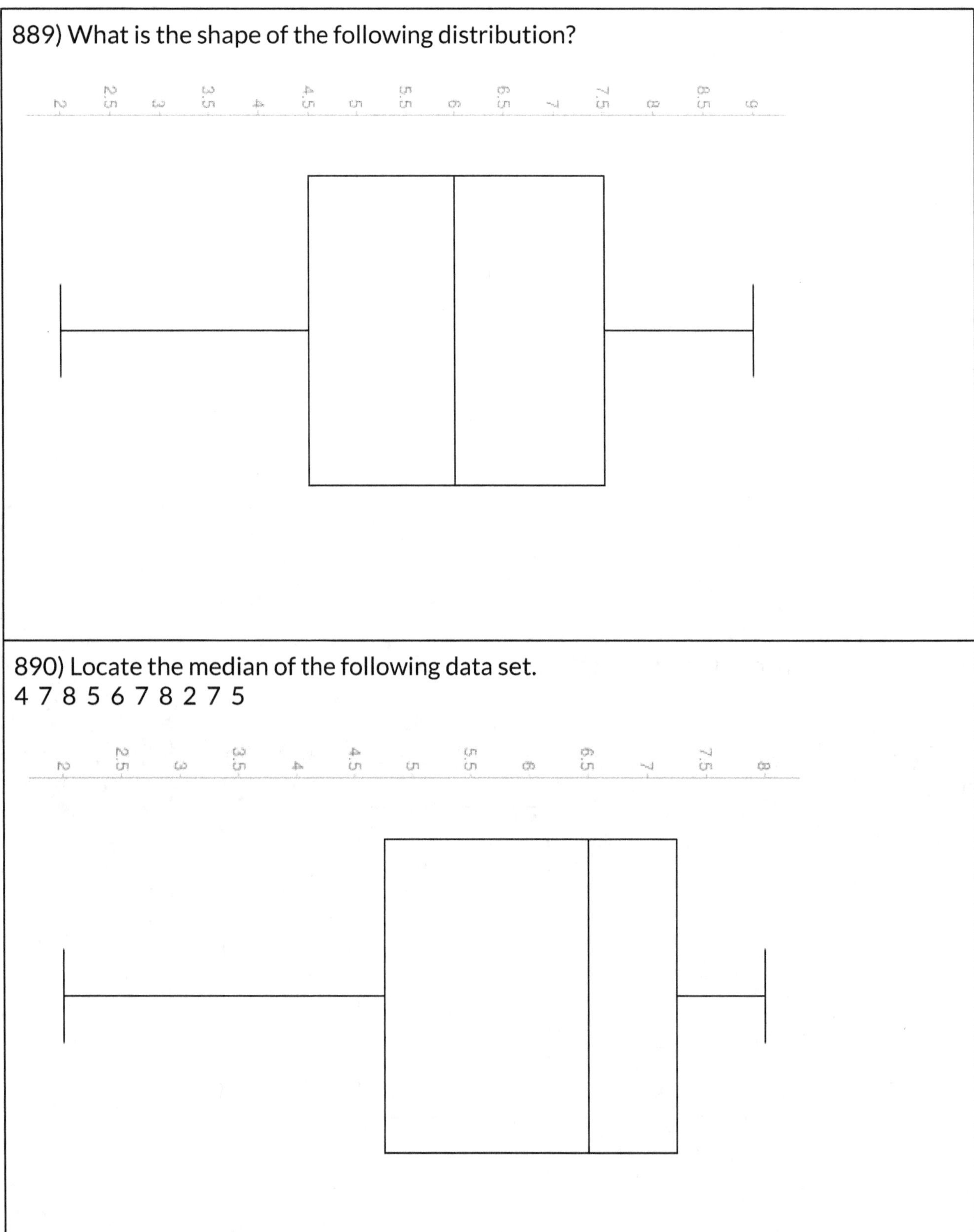

890) Locate the median of the following data set.
4 7 8 5 6 7 8 2 7 5

5.3 - Measure of Center and Variation

EXAMPLE: The following frequency table shows the number of hours of sleep each employee got on vacation. How many employees are there? What were the mean hours of sleep?

Number of hours of sleep	Number of employees
3	1
1	0
4	2
2	4
6	1

Solution: The total number of employees is $1 + 0 + 2 + 4 + 1 = 8$.

The mean is equal to the average. To get the mean, find the sum of all the *number of hours of sleep* and divide it by the total employees.

Mean = $3 + 4 + 4 + 2 + 2 + 2 + 2 + 6 = \frac{25}{8} = 3.125$ hours of sleep

Answer: **There are eight employees, and the mean hours of sleep is 3.125.**

EXAMPLE: What is the median of 4, 6, 1, 3, 0, 4?

Solution:

The *median* is the middle value of a given data set. If the number of data points is an even number, add the middle values, and divide that resultant value by 2 to get the median. Note to order the data set as 0, 1, 3, 4, 4, 6.

In this data set, the number of data points is even.

Median = $\frac{3+4}{2} = \frac{7}{2} = 3.5$

Answer: **3.5**

">

5.3 - Measure of Center and Variation

891) What is the average of 1, 6, 1, 2, 3, 7, 9?

892) What is the median of 4, 6, 1, 2, 3, 7, 3, 6?

5.3 - Measure of Center and Variation

893) What is the mean of 4, 6, 1, 3, 0, 4?

894) What is the mean of this group? 34, 19, 20, 45, 89, 9, 34

5.3 - Measure of Center and Variation

895) What is the median for this set of numbers: 34, 19, 20, 45, 89, 9, 34?

896) These are the football scores for this season: 19, 11, 11, 18, 19. What is the mean score?

5.3 - Measure of Center and Variation

897) The following frequency table shows the number of hours of sleep that each employee got on vacation.

Number of hours of sleep	Number of employees
3	1
1	0
4	4
2	4
8	1
6	1

What is the total number of employees there?

898) Sally is 11 years old. Samantha is also 11 years old. If Jacob is 15 years old, what would be the average age of them?

5.3 - Measure of Center and Variation

899) The following frequency table shows the number of hours of sleep that each employee got on vacation.

Number of hours of sleep	Number of employees
3	1
1	0
4	2
2	4
6	1
7	2
1	3

What is the total number of employees there?

900) John scored 79 on average in math. His average is based on seven tests. His scores on the first six tests are 90, 81, 70, 69, 81, 70. What score does John need to get in the 7th test to have an average of 79?

5.4 - Dot Plots, Frequency Tables, Histograms, and Box Plots

EXAMPLE: Jacob had different types of candies at different prices. Here each dot represents each candy. What is the maximum price of the candy?

Solution: From the above graph, see that the prices range from $1 to $5. The maximum price of the recorded candy is $5 each.

Answer: **$5**

EXAMPLE: The following frequency table shows the number of players that each group number of the soccer team has.

Number of Players	Number of Groups
2	Group 1
4	Group 2
5	Group 3
6	Group 4

Which member group has the lowest number of players?

Solution: From the frequency table, see that the lowest number of players are two, and it is in Group 1.

Answer: **Group 1**

5.4 - Dot Plots, Frequency Tables, Histograms, and Box Plots

901) Jacob had different types of candies at different prices. Here each dot represents each candy.

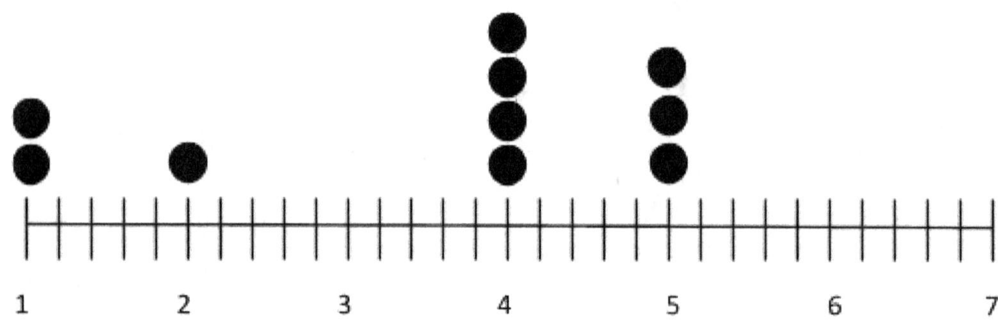

How many candies cost $5 each?

902) Jacob had different types of candies at different prices. Here each dot represents each candy.

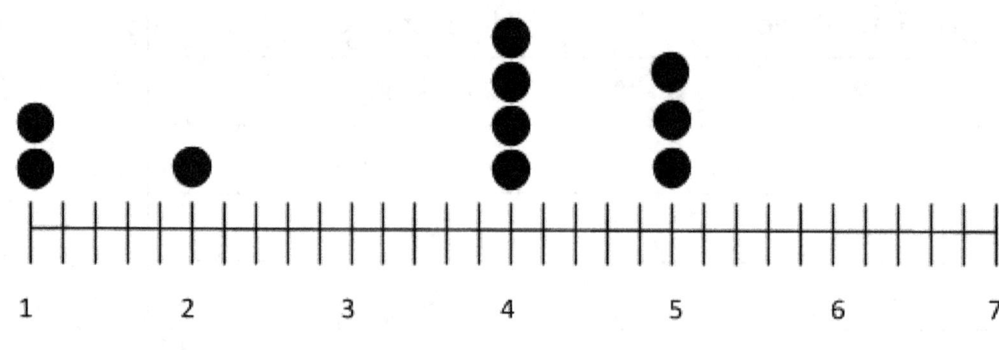

What is the total number of candies that cost $1 and $4 each?

5.4 - Dot Plots, Frequency Tables, Histograms, and Box Plots

903) The following dot plot shows the number of problems done by each student given by the instructor last week. Each dot represents a different student.

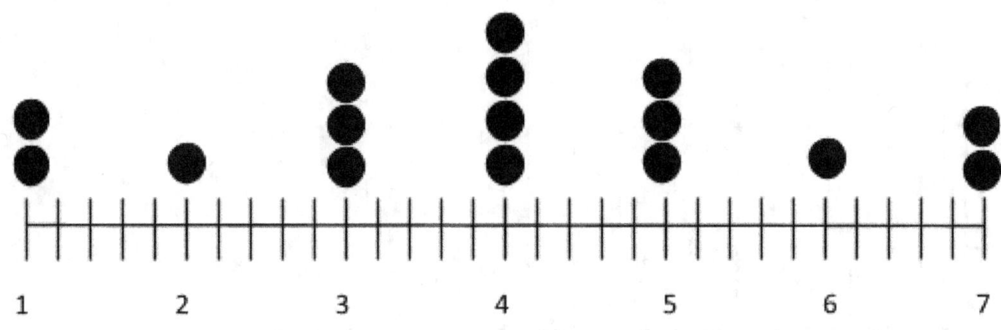

What was the minimum number of problems that a student had?

904) The following dot plot shows the number of problems done by each student given by the instructor last week. Each dot represents a different student.

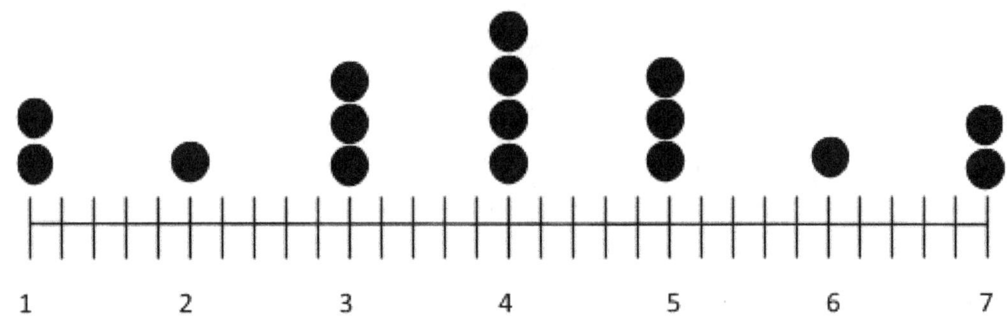

What was the maximum number of problems that a student had?

5.4 - Dot Plots, Frequency Tables, Histograms, and Box Plots

905) Sanda was hospitalized. The following line diagram shows her temperature in one day at an interval of 3 hours.

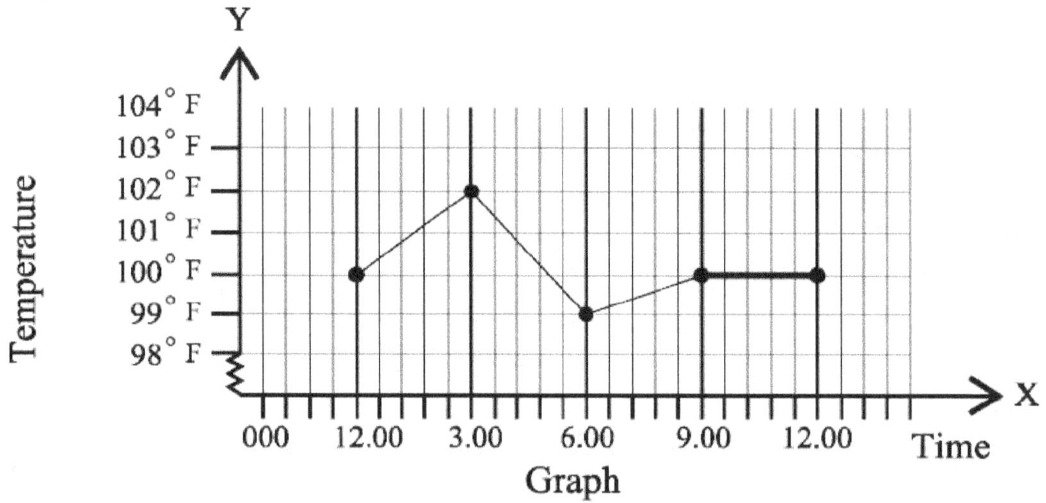

What was the maximum temperature Sanda experienced?

906) Sanda was hospitalized. The following line diagram shows her temperature in one day at an interval of 3 hours.

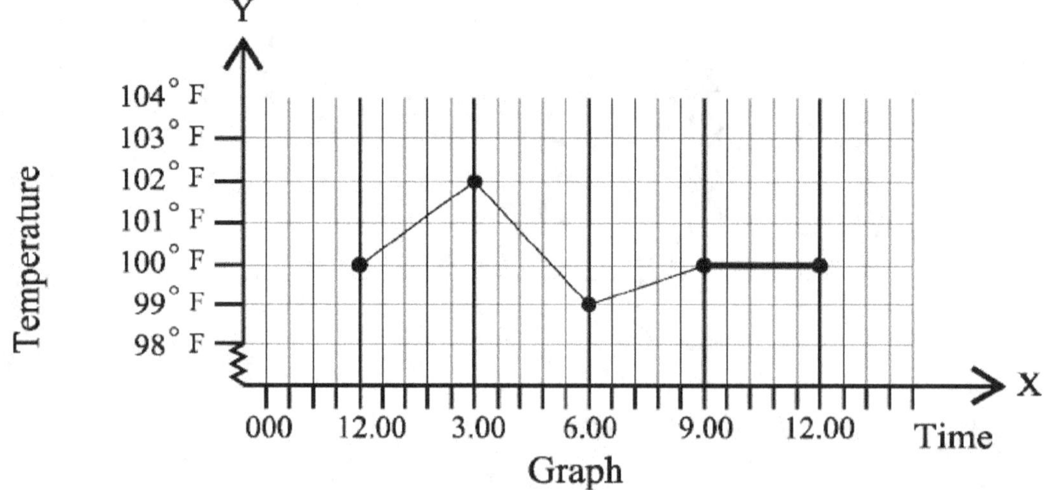

What was the temperature Sanda had at noon?

5.4 - Dot Plots, Frequency Tables, Histograms, and Box Plots

907) The histogram shows the weight of students in a 6th-grade class.

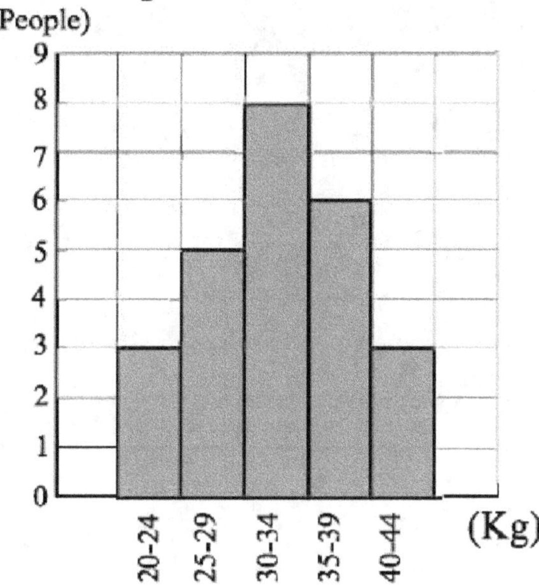

Which weight group has the most students?

908) The histogram shows the weight of students in a 6th-grade class.

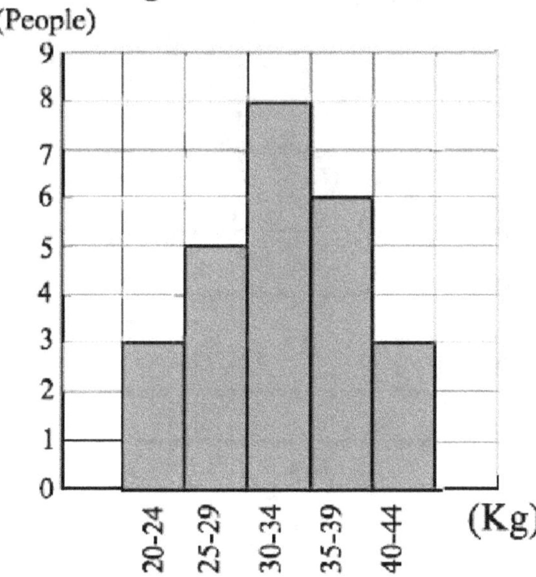

How many students have the weights in 35-39 (kg)?

5.4 - Dot Plots, Frequency Tables, Histograms, and Box Plots

909) The histogram shows the weight of students in a 6th-grade class.

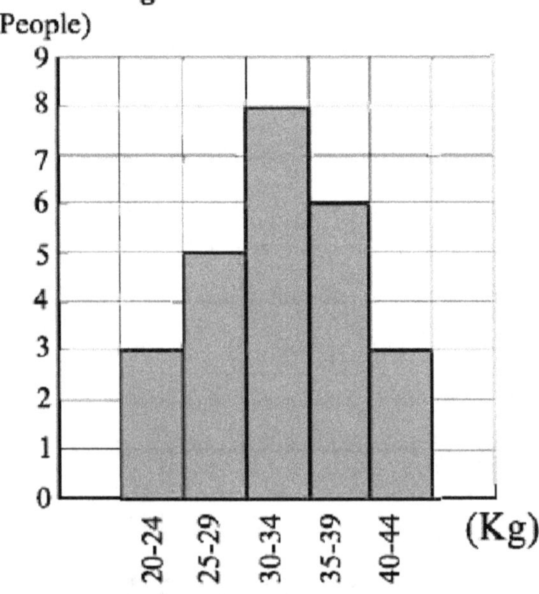

How many students are in the class?

910) The following dot plot shows how the number of times that section A students submitted their assignments are distributed. Each dot represents one student.

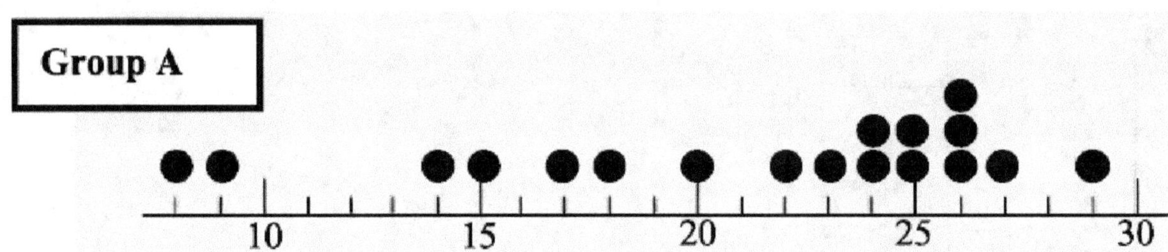

What is the total number of students in the class?

5.4 - Dot Plots, Frequency Tables, Histograms, and Box Plots

911) In a winter season in New York City, the temperature (in celsius) of September was as follows:

14°, 14°, 14°, 13°, 12°, 13°, 10°, 10°, 11°, 12°, 11°, 10°, 9°, 8°, 9°, 11°, 10°, 10°.

What was the range of temperature?

912) In a winter season in New York City, the temperature (in celsius) of September was shown as follows:

14°, 14°, 13°, 12°, 13°, 10°, 10°, 11°, 12°, 11°, 10°, 9°, 8°, 9°, 11°, 10°, 10°.

What was the mean temperature in September?

5.4 - Dot Plots, Frequency Tables, Histograms, and Box Plots

913) Emily notices how many times he ate an apple each month. She ate apples ten times in August, 12 times in March, ten times in July, and five times in December. What is her median number of times eating apples?

914) Below is Sam's test scores in different subjects out of 30 items.

English	Math	Geography	Philosophy	Biology
29	18.5	21.5	26	24

What is Sam's mean test score?

5.4 - Dot Plots, Frequency Tables, Histograms, and Box Plots

915) The following frequency table shows the number of players each group number of the soccer team has.

Number of Players	Number of Groups
2	Group 1
4	Group 2
4	Group 3
6	Group 4

What is the median number of players?

916) The following frequency table shows the number of players each group number of the soccer team has.

Number of Players	Number of Groups
2	Group 1
4	Group 2
5	Group 3
6	Group 4

Which member group has the highest number of players?

5.4 - Dot Plots, Frequency Tables, Histograms, and Box Plots

917) The following frequency table shows the number of people that each jury board has.

Number of People	Name of Jury Board
2	Delta
6	Gamma
4	Kappa
6	Sigma

What is the median number of people?

918) What is the median of the following box plot?

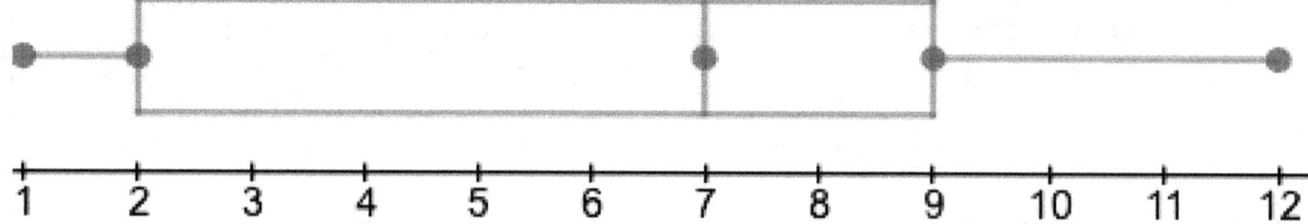

5.4 - Dot Plots, Frequency Tables, Histograms, and Box Plots

919) What is the interquartile of the following box plot?

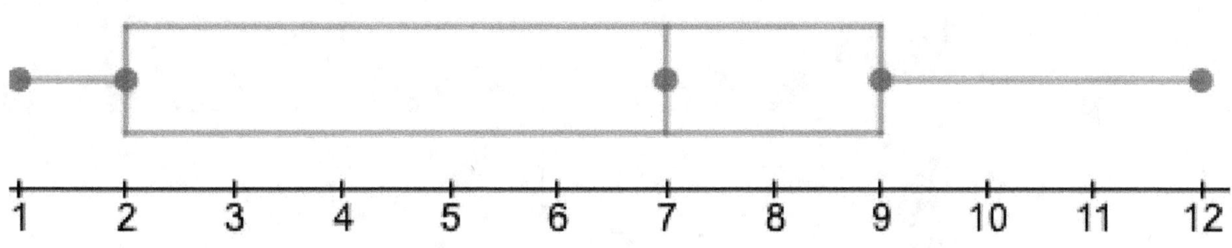

920) What is the range of the following box plot?

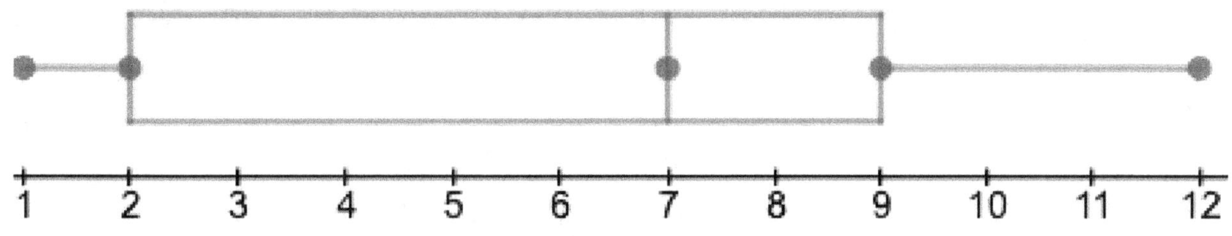

5.5A - Interpret Histograms and Dot Plots

A *histogram* is a graphical representation of data and values using bars of different heights. A *dot plot* is a graphical representation of data using dots.

EXAMPLE: What is the maximum y-value of this histogram?

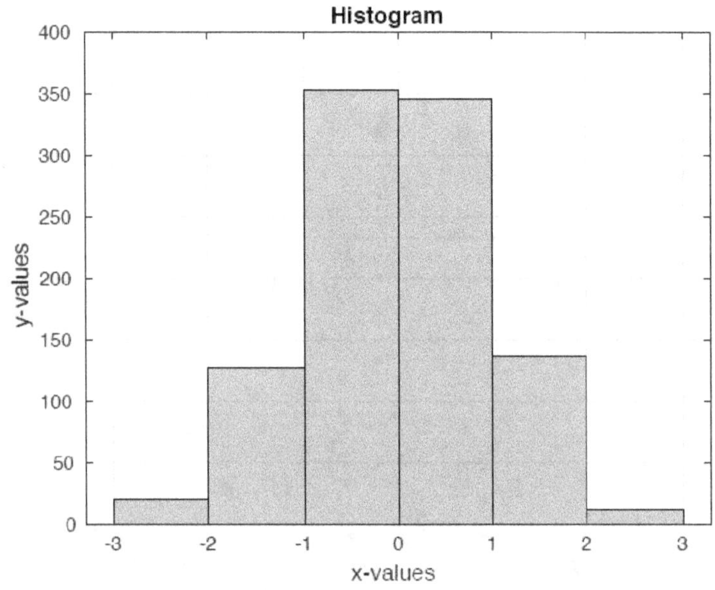

Answer: The highest bar indicates a value of 350 at y-values. Therefore, the maximum y-value is 350.

EXAMPLE: How many candies cost $7?

Sarah had different types of candies at different prices. Each dot on the plot indicates one candy.

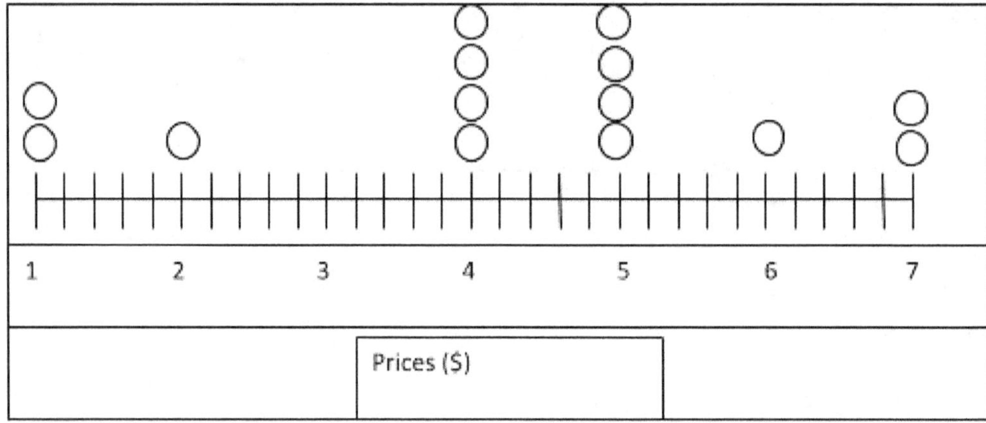

From the dot plot, two candies are at $1, one candy at $2, four candies at $4, four candies at $5, one candy at $6, and two candies at $7.

Answer: **2 candies**

5.5A - Interpret Histograms and Dot Plots

921) The following histogram shows the changes in the number of buildings in the area "A" from 1990 to 2020.

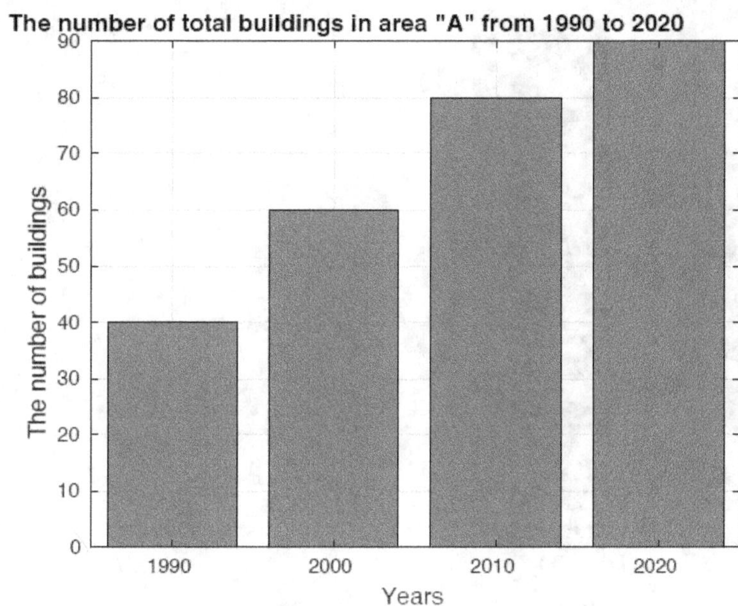

In which year the number of buildings was the least?

922) Liana went to a doll's shop with her father. Her father got her three dolls.

How much did Liana's father need to pay?

5.5A - Interpret Histograms and Dot Plots

923) Find the total number of detective books.

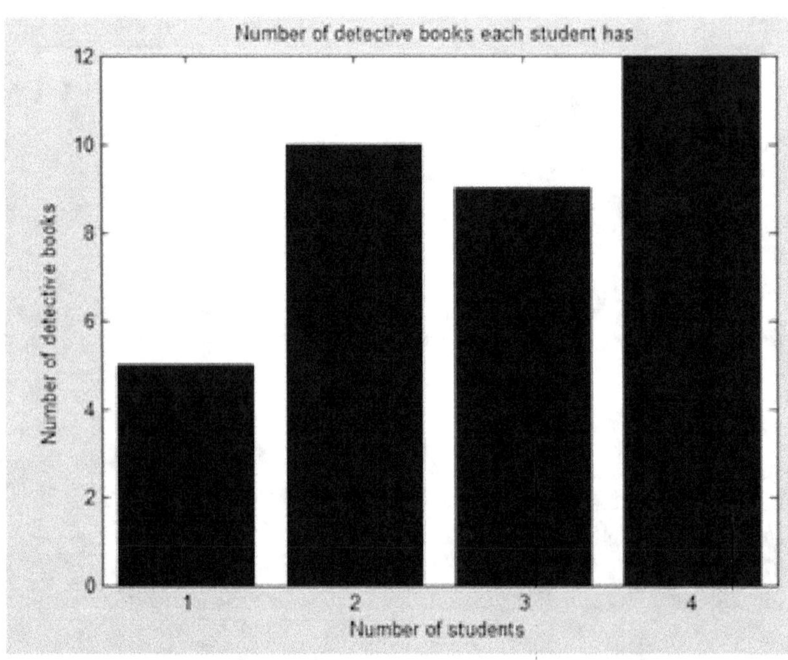

924) How many students do have exactly nine detective books?

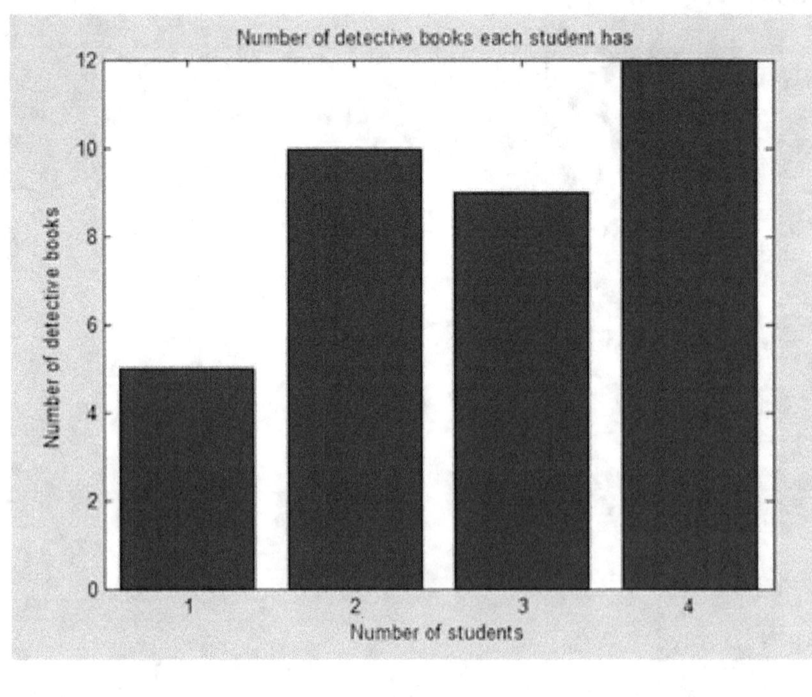

5.5A - Interpret Histograms and Dot Plots

925) The following histogram represents the number of children in each family in any specific area. Using this histogram, find the number of families where there is one child.

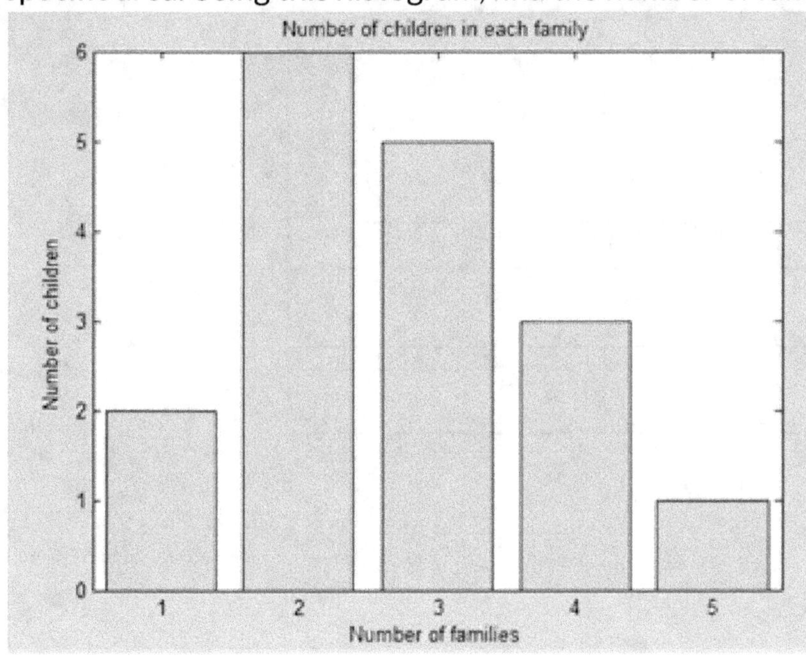

926) The following histogram represents the number of children in each family in any specific area. Using this histogram, find the total number of children in that area.

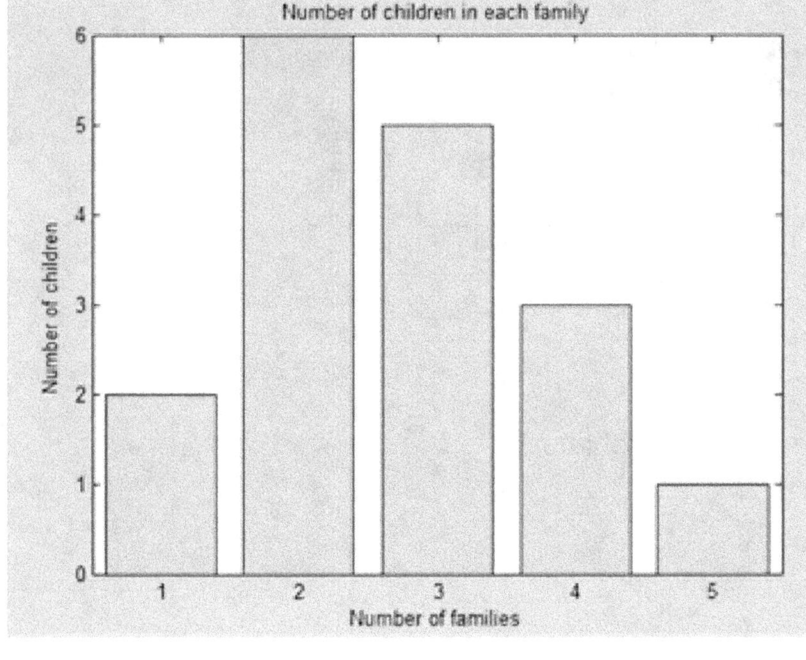

5.5A - Interpret Histograms and Dot Plots

927) Wendy had different types of candies of different prices. Each dot represents each candy.

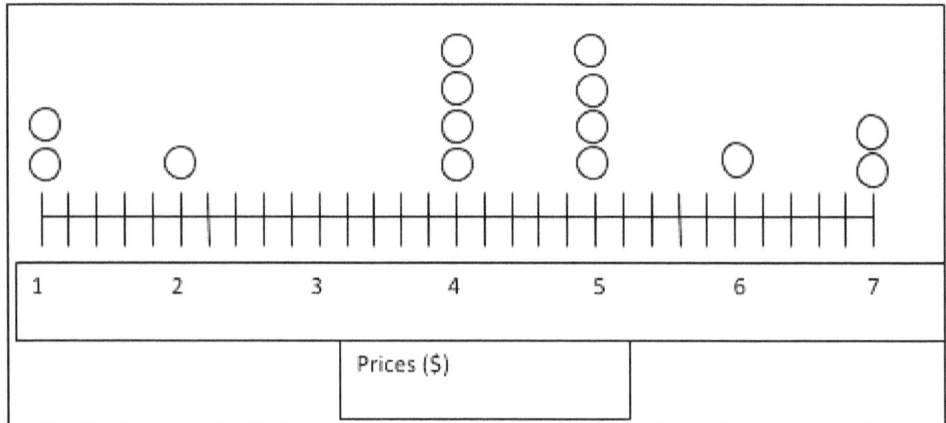

How many candies cost $5?

928) There are a total of seven problems in an assignment. Each dot represents each student.

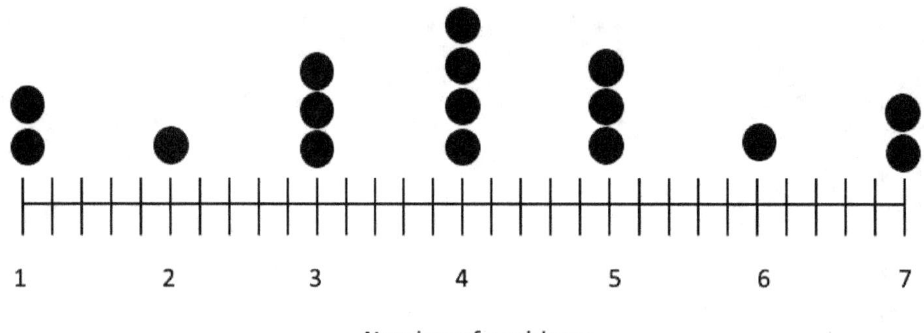

How many students could solve all seven problems?

5.5A - Interpret Histograms and Dot Plots

929) There are a total of seven problems in an assignment. Each dot represents each student.

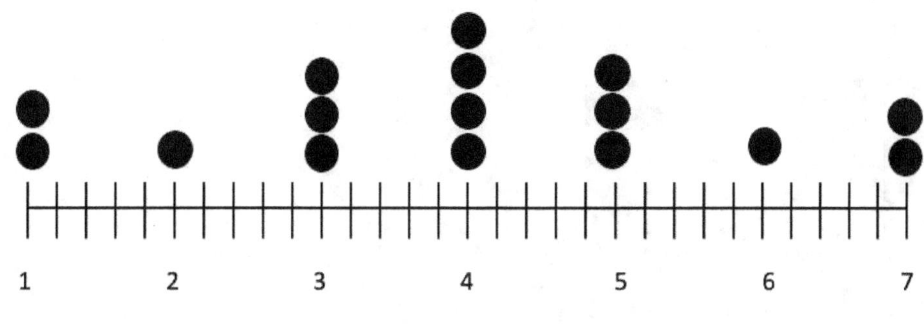

How many students could solve only one problem?

930) In a library, the number of books (special edition) is shown in the following plot. Here the corresponding prices are for each of the books

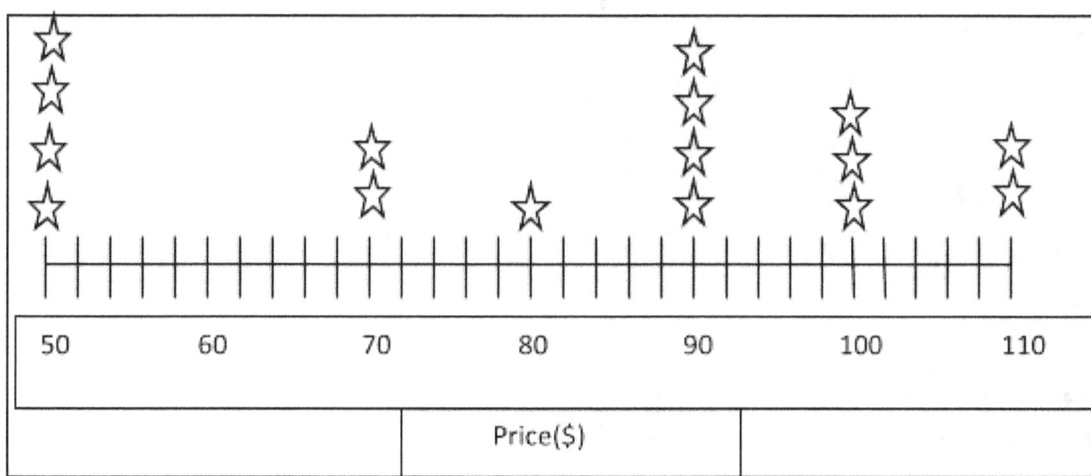

How many total books are there in the library?

5.5B - Samples

EXAMPLE: The owner of a flower shop counted the total flowers that she sold last month.

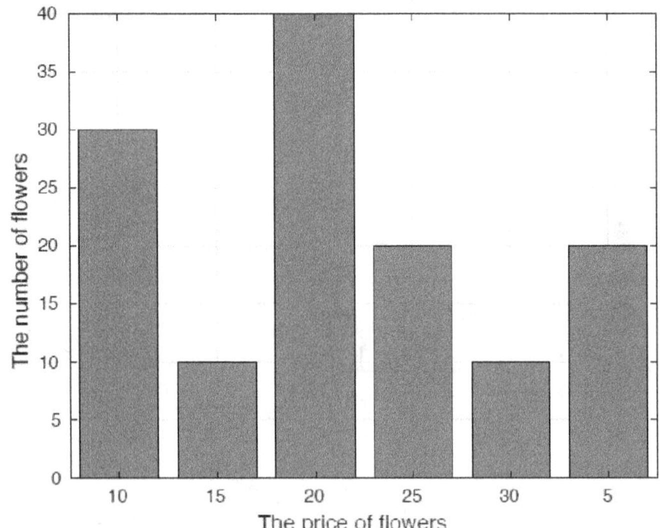

How many flowers did she sell at $20?

Solution: The bar at $20 reaches the number of flowers at 40. Therefore, she sold 40 flowers at $20.

Answer: **The owner sold 40 flowers at $20.**

EXAMPLE: The following histogram shows the changes in the number of buildings in the area "A" from 1990 to 2020. How many buildings are there in 2010?

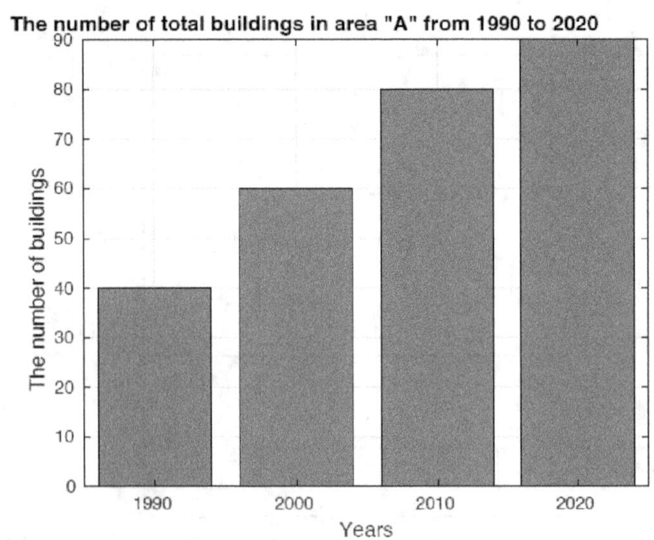

Solution: The bar in 2010 reached the number of buildings at 80. Therefore, there were 80 buildings in 2010.

Answer: **There were 80 buildings in 2010.**

5.5B - Samples

931) The following histogram shows the changes in the number of buildings in the area "A" from 1990 to 2020.

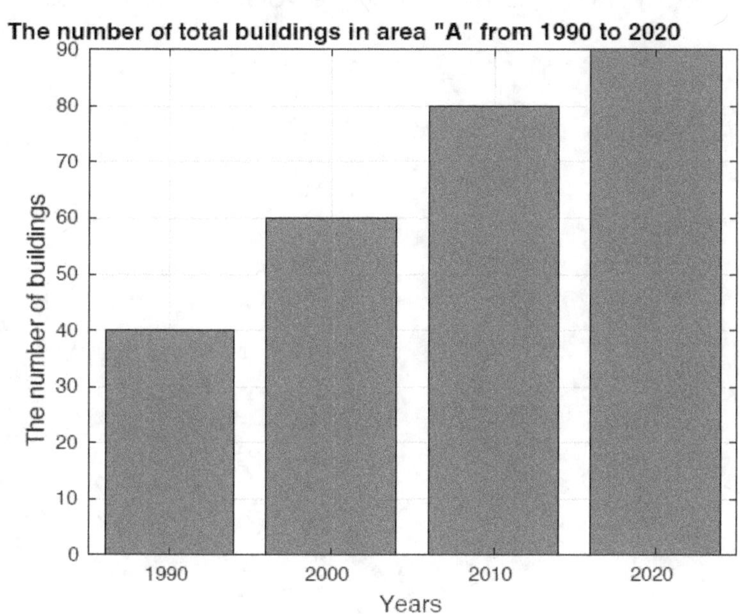

What is the range of this histogram?

932) The following histogram shows the changes in the number of buildings in the area "A" from 1990 to 2020.

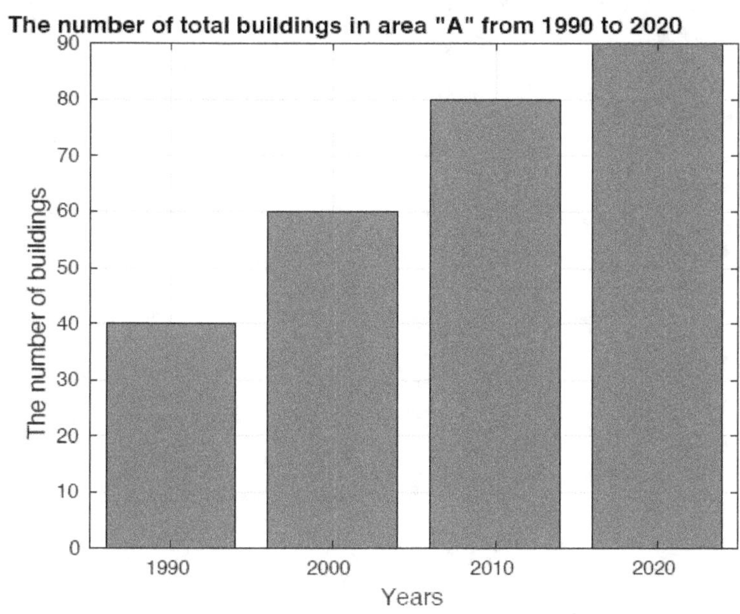

In which year the number of buildings was the maximum?

5.5B - Samples

933) Belinda was tasked to solve some homework problems at the end of the last week. Each problem weighs equally. To find out the difficulty, she chose three problems from each of the three sections of the homework. Is the sample of homework problems likely to be representative?

934) As a psychology student, David studied four different real-life criminal cases. Each criminal came from the same type of background and committed an equal number of crimes. Is the sample of criminal cases likely to be biased?

5.5B - Samples

935) As a psychology student, Jessica had to study some different cases to complete her course work. She studied four domestic violence cases. Is the sample of cases likely to be biased?

936) The school authority surveyed the 15 senior teachers in the school. Is the sample of the teachers in the school likely to be representative?

5.5B - Samples

937) The owner of a flower shop counted the total flowers that she sold last month.

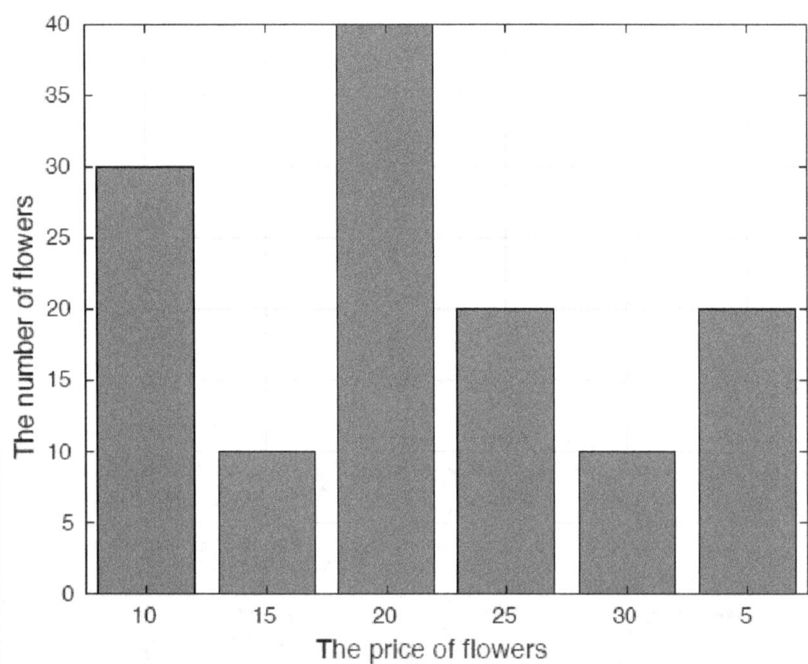

How many flowers did she sell last month?

938) The following histogram shows the changes in the number of buildings in the area "A" from 1990 to 2020.

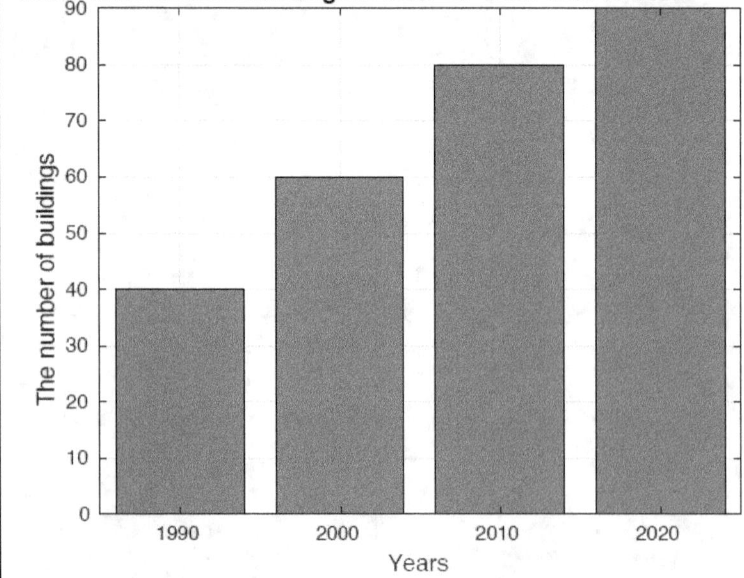

How many buildings are there in 2020?

5.5B - Samples

939) The following histogram shows the changes in the number of buildings in the area "A" from 1990 to 2020.

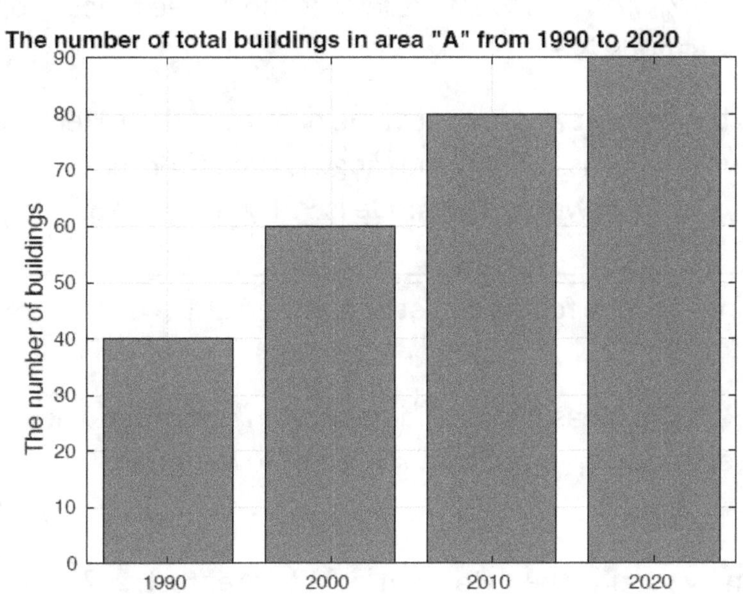

What is the median number of buildings in the area "A"?

940) Sarah had different types of candies of different prices. Each dot represents each candy.

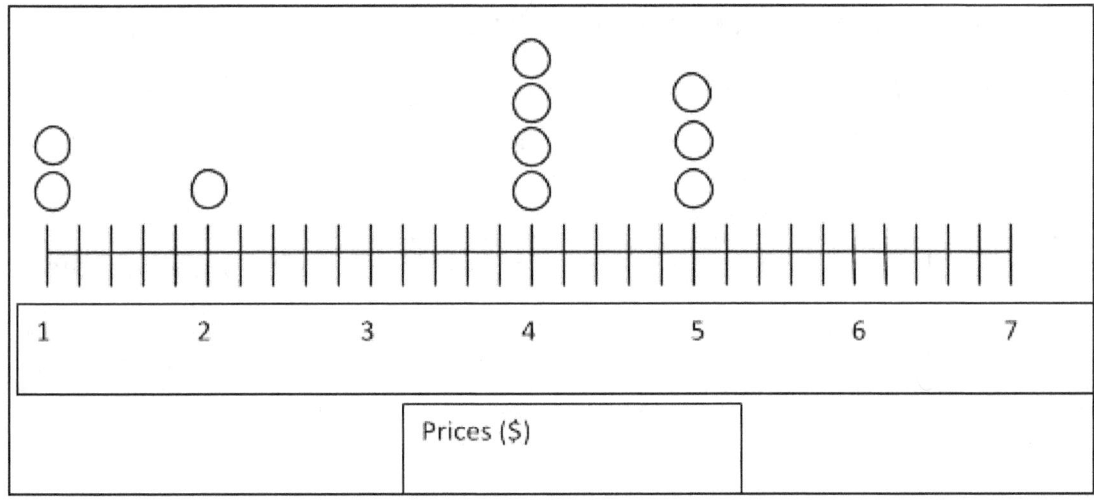

What's the total number of Sarah's candies?

5.5C - Mean, Median, Mode, Range, Quartiles, IQR, and MAD

EXAMPLE: Find the mode and the range of the following list of numbers.
2, 3, 8, 2, 5, 7, 9, 6, 7, 8, 9, 10, 3, 4, 9

Solution: The *mode* is the most occurred number of a data set. To identify the mode, order the data set as 2, 2, 3, 3, 4, 5, 6, 7, 7, 8, 8, 9, 9, 9, 10. Here, 9 occurs three times and is the most repeated number. Thus, the mode is 9.

The *range* is the difference between the highest and lowest value of a set of data. Here, the highest value is 10, while the lowest value is 2. Thus, the range is the difference between 10 and 2, which is 8.   Answer: **The mode is 9. The range is 8.**

EXAMPLE: Find the interquartile range for the following list of numbers.
2, 7, 9, 10, 8, 12, 11

Solution: The *interquartile range (IQR)* is the measure of "middle fifty" in a data set. The IQR is measured by the difference between upper quartile (Q_3) and lower quartile (Q_1). Thus, $IQR = Q_3 - Q_1$.

To find the quartiles, find the median by sorting the data from least to greatest. 2, 7, 8, **9**, 10, 11, 12. The median is **9**. The lower quartile (Q_1) is the median of the data to the *left of the median*. 2, **7**, 9. The lower quartile is **7**. The upper quartile (Q_3) is the median of the data to the *right of the median*. 10, **11**, 12. The upper quartile is **11**. Hence, $IQR = Q_3 - Q_1 = 11 - 7 = 4.$ Answer: **The interquartile range is 4.**

EXAMPLE: Find the mean absolute deviation (MAD) of 2, 4, 8, 6.

Solution: The *mean absolute deviation (MAD)* of a data set is the average distance between each data value and the mean. To find MAD, find the mean first. The mean is $\frac{2+4+8+6}{4} = \frac{20}{4} = 5$. Then, find the distance of each data value from the mean.

Data value	Distance from the mean	Data value	Distance from the mean
2	5 - 2 = 3	8	8 - 5 = 3
4	5 - 4 = 1	6	6 - 5 = 1

The MAD is the average of the distances from the mean, $\frac{3+1+3+1}{4} = \frac{8}{4} = 2$.

Answer: **The mean absolute deviation is 2.**

5.5C - Mean, Median, Mode, Range, Quartiles, IQR, and MAD

941) Find the mode for the following list of numbers. 2, 3, 8, 2, 5, 7, 9, 6	942) What is the mode of the following set of numbers? 5 , 9, 6, 2, 5, 8, 5
943) Find the mode of the list of numbers. 11, 4, 2, 7, 23, 9, 2, 5, 2, 11	944) Find the mean for the following list of numbers. 2, 7, 9, 10
945) What is the mean of the following set of numbers? 7, 9, 6, 2, 5, 8, 5	946) What is the mean of the following set of numbers? 120, 43, 98, 64, 23, 45, 67, 12
947) Find the median for the following list of numbers. 2, 7, 9, 10, 8	948) What is the median of the following set of numbers? 5, 8, 5, 7, 9, 6, 20, 12, 23
949) What is the median of the following set of numbers? 78, 45, 78, 23, 45, 67, 90, 12, 35	950) Find the range for the following list of numbers. 2, 7, 9, 10, 8

5.5C - Mean, Median, Mode, Range, Quartiles, IQR, and MAD

951) What is the range of the following set of numbers? 78, 45, 78, 23, 45, 67, 90, 12, 35	952) What is the range of the following set of numbers? 5, 8, 5, 7, 9, 6, 20, 12, 23
953) What is the range of the following set of numbers? 2300, 4567, 8902, 3456, 8732, 7654	954) Find the upper quartile of the given set of numbers. 13, 17, 11, 12, 5, 9, 4
955) Find the lower quartile of the given set of numbers. 13, 17, 11, 12, 5, 9, 4	956) Find the middle quartile of the given set of numbers. 13, 17, 11, 12, 5, 9, 4
957) Find the lower quartile of the given set of numbers. 14, 17, 21, 18, 24, 18, 7, 3, 10, 4	958) Determine the interquartile range of the following set of numbers. 13, 17, 11, 12, 5, 9, 4
959) Determine the interquartile range of the following set of numbers. 14, 17, 21, 18, 24, 18, 7, 3, 10, 4	960) Calculate the interquartile range of the following set of numbers. 132, 120, 147, 310, 250, 77, 90

5.5C - Mean, Median, Mode, Range, Quartiles, IQR, and MAD

961) What is the median of the set of numbers: 14, 17, 21, 18, 24, 18, 7, 3, 10, 4?	962) What is the mode of the set of numbers: 14, 17, 21, 18, 24, 18, 7, 3, 10, 2?
963) Here are the numbers of flower bouquets that Sosa sold last week. 33, 24, 15, 19, 37, 33, 40 Find the median of the numbers of flower bouquets.	964) Here are the numbers of flower bouquets that Riana sold last week. 33, 24, 15, 19, 37, 33, 40 Find the interquartile range of the numbers of flower bouquets.
965) Calculate the median of the given set. 7, 8, 3, 14, 7, 9, 7	966) Calculate the median of the given set. 37, 8, 3, 24, 7, 49, 7, 9, 7
967) Samantha works as a nanny who has experience in taking care of children of different ages. 5, 9, 2, 1, 11, 8, 4, 9, 6 are the ages of the children she has taken care of so far. What is the mode of the set of ages?	968) Paul works as a babysitter who has experience in taking care of children of different ages. 5, 9, 2, 1, 11, 8, 4, 9, 6 are the ages of the children she has taken care of so far. What is the median of the set of ages?
969) Calculate the median of the given set. 37, 8, 3, 24, 7, 49, 7	970) Find the mode of the given set. 37, 8, 3, 24, 24, 7, 49, 7

5.5C - Mean, Median, Mode, Range, Quartiles, IQR, and MAD

971) The following frequency table has been made by collecting data from a survey. The survey was conducted on the favorite TV channels of 250 people. How many people like to watch ABC?

TV Channel	Fox	ABC	CNN
Frequency	45	75	20

972) The following frequency table has been made by collecting data from a survey. The survey was conducted on the favorite TV channels of 250 people. Which one is the most popular channel?

TV Channel	Fox	ABC	CBS
Frequency	45	75	110

5.5C - Mean, Median, Mode, Range, Quartiles, IQR, and MAD

973) The following graph represents the number of customers that the customer care center of a mobile company served last week.

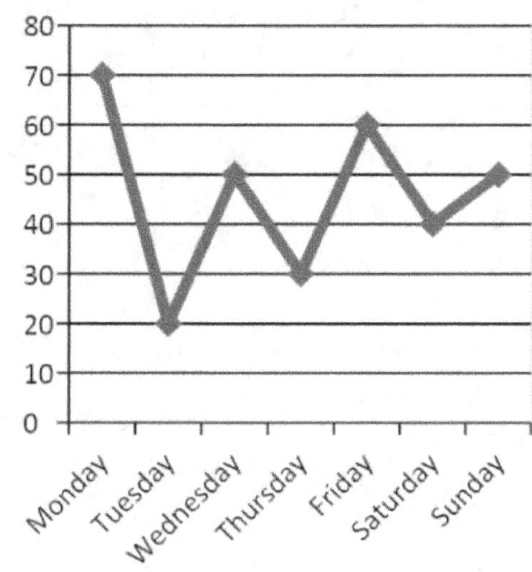

How many customers could they serve last week?

974) The following graph represents the number of customers the customer care center of a mobile company served last week.

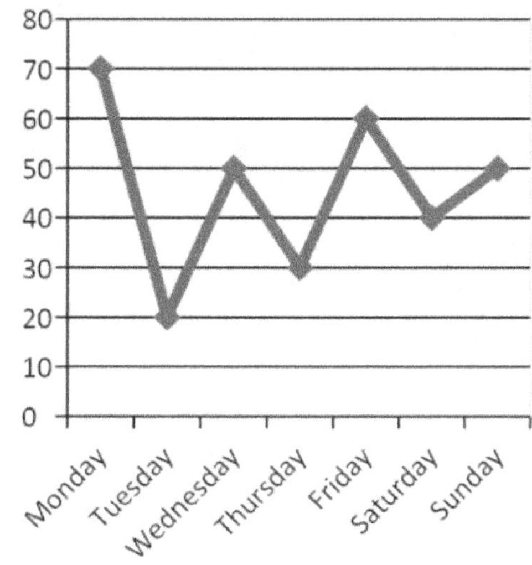

What is the range of the numbers of customers?

5.5C - Mean, Median, Mode, Range, Quartiles, IQR, and MAD

975) The following graph represents the number of books in a library.

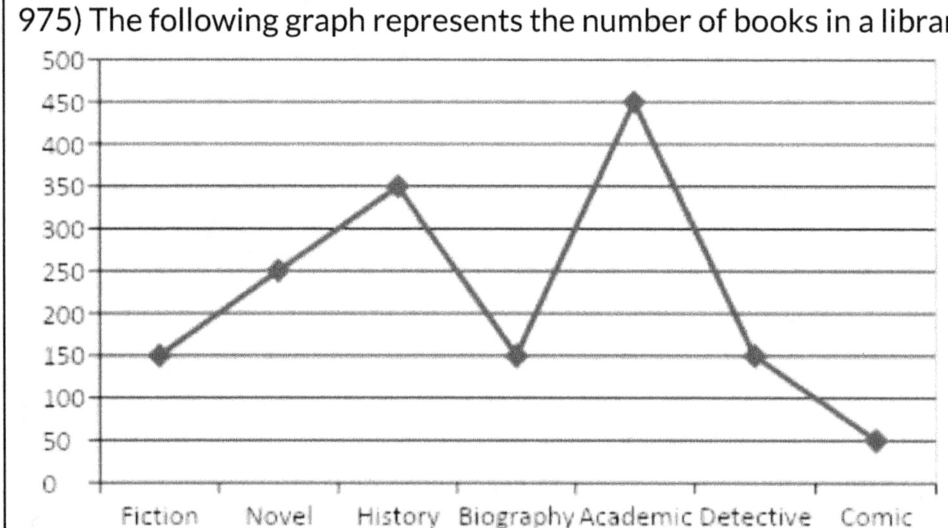

What is the mode of the numbers of books?

976) The following graph represents the number of books in a library.

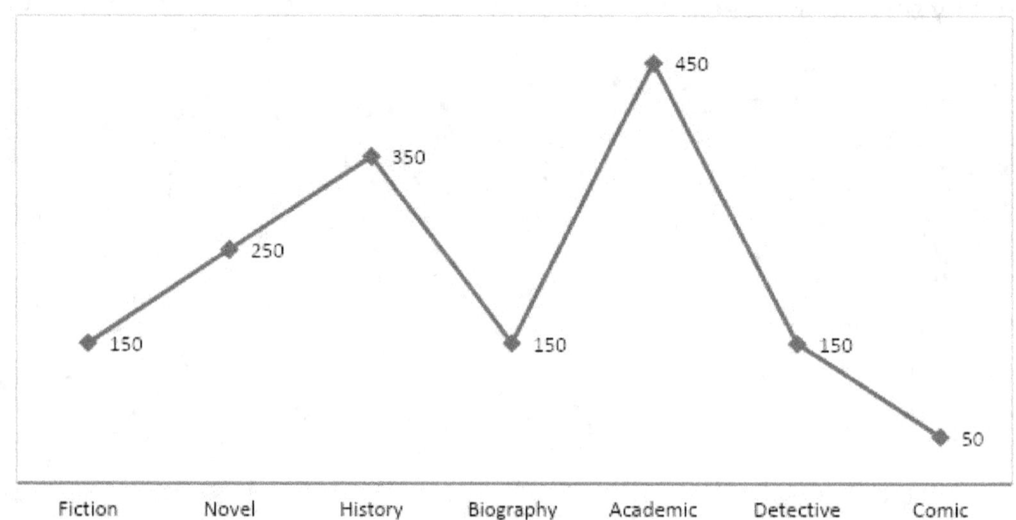

What is the range of the numbers of books?

5.5C - Mean, Median, Mode, Range, Quartiles, IQR, and MAD

977) The following graph represents the number of flowers from Sarah's garden on weekdays.

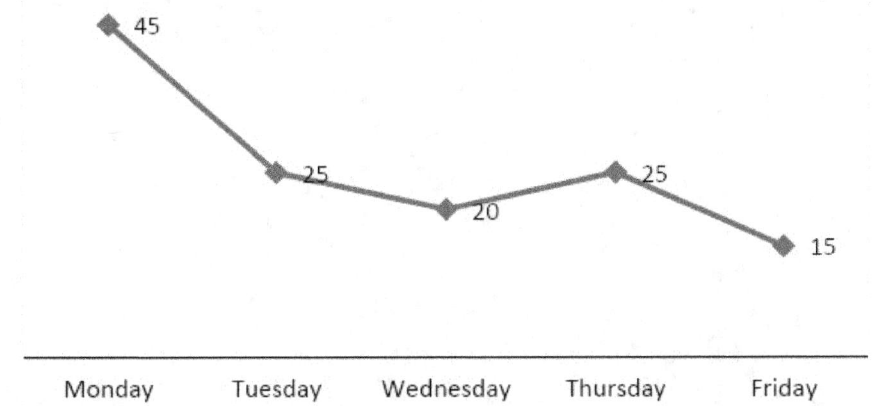

What is the range of the number of flowers?

978) The following graph represents the number of flowers from Sarah's garden on weekdays.

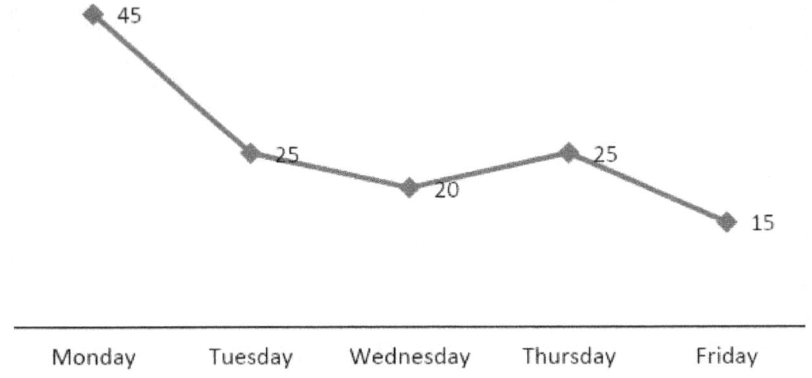

What is the mode of the number of flowers?

5.5C - Mean, Median, Mode, Range, Quartiles, IQR, and MAD

979) If the outlier is removed from the set 2, 5, 8, 4, 20, 150, 3, what would happen to the mean of the set of the numbers?

980) If the outlier is removed from the set 250, 560, 780, 230, 5, 450, 780, 950, what would happen to the mean of the set of the numbers?

5.5C - Mean, Median, Mode, Range, Quartiles, IQR, and MAD

981) Determine the lower quartile of 2, 5, 8, 4, 20, 15, 3.

982) Determine the upper quartile of 2, 5, 8, 4, 20, 15, 3.

5.5C - Mean, Median, Mode, Range, Quartiles, IQR, and MAD

983) Calculate the mean absolute deviation (MAD) of the set of numbers: 4, 7, 8, 9.

984) If the mean of the following set of numbers is 59, what is the missing number?
120, 43, 98, 64, 23, _______, 67, 12

5.5C - Mean, Median, Mode, Range, Quartiles, IQR, and MAD

---

985) If the mean of the following set of numbers is 43.8, what is the missing number?

_______, 67, 12, 23, 85

---

986) If the mean of the following set of numbers is 59, what is the missing number?

120, 43, 66, 64, 23, _______, 67, 12

5.5C - Mean, Median, Mode, Range, Quartiles, IQR, and MAD

987) If the mean of the following set of numbers is 60, what is the missing number?
120, 43, 98, 64, 23, _______, 67, 12

---

988) The following is the five-number summary of the numbers of students at five different schools in an area.

Minimum	Lower Quartile	Median	Upper Quartile	Maximum
240	320	450	480	530

What percent of the schools has more than 480 students?

5.5C - Mean, Median, Mode, Range, Quartiles, IQR, and MAD

989) The following is the five-number summary of the numbers of students at five different schools in an area.

Minimum	Lower Quartile	Median	Upper Quartile	Maximum
240	320	450	480	530

What percent of the schools has less than 480 students?

990) The median of the number of pens that five friends have is 4. Everyone has at least one pen. If the range of the number is 5, then find the validity of the following statement.

'The highest number of pens that one has is 4'.

5.5D - Data Word Problems

EXAMPLE: A group of four students likes to play cricket together, and each student keeps track of his all-time maximum score in a single play. Their high scores are all between 120 and 150, except for Jake, whose maximum score is 220. Jake then played a great game and got a new maximum score of 310. How does Jake's new score affect the median?

Solution: Jake's new maximum score of 310 will not affect the median of the score set, because the data scores necessary to calculate the median are the two scores at the middle of the set, which are still lower than Jake's new maximum score.

Answer: **There will be no change in the median.**

EXAMPLE: Fiona works at an aquarium store. One day, she was looking at some data on the weights of five fishes. The mean weight of the fishes was 2 lbs, and the median was 1 lb. The heaviest fish, a catfish, was recorded as 3 lbs.

Fiona then discovered that the catfish's weight was miswritten. The actual weight should have been written as 4 lbs. How does the new weight of catfish affect the mean and median?

Solution: Increasing the catfish's weight to 4 lbs will significantly increase the mean, because the total weight of the fishes will increase significantly. However, increasing catfish's weight will not affect the median, because the median, which is 1 lb, is still lower than the catfish's new weight.

Answer: **The mean will increase, and the median will remain the same.**

EXAMPLE: A group of four students likes to play cricket together, and each student keeps track of his all-time maximum score in a single play. Their high scores are all between 120 and 150, except for Jake, whose maximum score is 220. Jake then played a great game and got a new maximum score of 310. How does Jake's new score affect the mean?

Solution: Increasing Jake's maximum score to 310 will significantly increase the mean, because the total maximum score of the students will increase significantly.

Answer: **Jake's new score will change the mean of the data. The mean will increase.**

5.5D - Data Word Problems

<table>
<tr><td>

991) Fiona received her five assignment scores. Then, she scored for another course, which is 18. The following is an illustration of her scores.

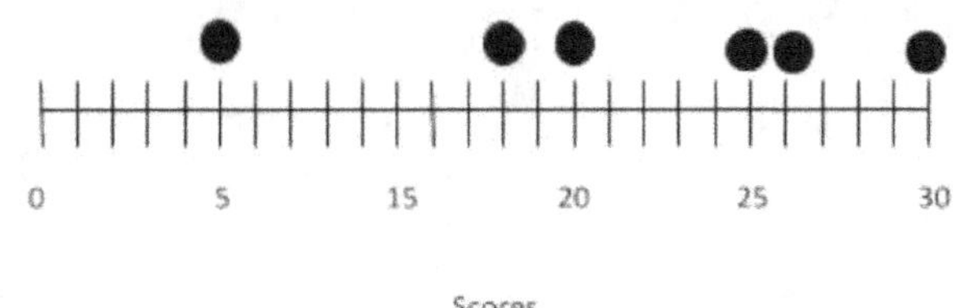

What did happen to the mean of the new set of the assignment scores?

</td></tr>
<tr><td>

992) Rosa received her five assignment scores. Then, she scored for another course, which is 18 . The following is an illustration of her scores.

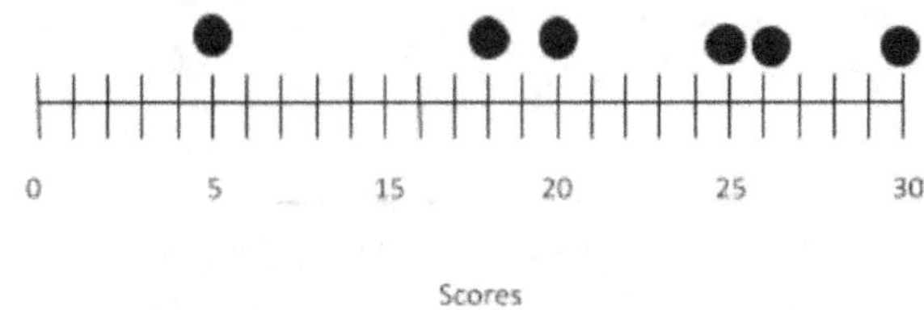

What did happen to the median of the new set of assignment scores?

</td></tr>
<tr><td>

993) A survey was conducted on the height of students in a school. After several days, a new student transferred in whose height is 164 cm. If earlier, the mean of the height of the students is 162.4 cm, what would happen to the new mean of the height of the students?

</td></tr>
<tr><td>

994) A survey was conducted on the height of students in a school. After several days, a new student transferred in whose height is 162 cm. If earlier, the mean of the height of the students is 162.4 cm, what would happen to the new median of the height of the students?

</td></tr>
<tr><td>

995) In a fruit basket, there were four large-sized fruits. Three fruits weigh between 80 grams to 100 grams. If Jennifer ate one of the fruits that weighed 120 grams, what would happen to the new mean?

</td></tr>
</table>

5.5D - Data Word Problems

996) In a fruit basket, there were four large-sized fruits. Three fruits weigh between 80 grams to 100 grams. If Jennifer ate one of the fruits that weighed 120 grams, would the median remain the same?

997) Jacob's father, mother, and younger sister used to earn $2,500, $1,800, and $2,000, respectively. Yesterday, Jacob started a job and will be earning $1,500. How would the mean of their earnings be affected?

998) Jacob's father, mother, and younger sister used to earn $2,500, $1,800, and $2000, respectively. Yesterday, Jacob started a job and will be earning $1,500. How would the median of their earnings be affected?

999) There were 24 birds in a shop for $27,000. The shop owner decided to enrich his bird's collection, bringing seven new birds each for $1,050. What would happen to the mean of the prices of the birds?

1,000) There were 24 birds in a shop for $27,000. The shop owner decided to enrich his bird's collection, bringing seven new birds each for $1,050. Determine whether the median would remain the same.

Answer Key -- 6th Grade Math Workbook

1. 2:3
2. 12:30 or 2:5
3. 54:12 or 9:2
4. 15:12 or 5:4
5. 45:10 or 9:2
6. 32:23
7. 125:5 or 25:1
8. 200:5 or 40:1
9. 140:125 or 28:25
10. 7:6
11. 3:1
12. 9:49
13. 1:2
14. 2:1
15. 3:1
16. 6:7
17. 32:33
18. 19:12
19. 10:9
20. 23:9
21. 2:3
22. 1:5
23. 1:30
24. 5:4
25. 5:7
26. 9:17
27. 8:13
28. 5:1
29. 25:81
30. 4:9
31. 30 students per classroom
32. 20 students per bus
33. 3 ice creams per kid
34. 25 fruits per basket
35. 13 copies per min
36. $32 per hour
37. 2 pencils per student
38. 2 grams of water per gram of vinegar
39. 25 books per shelf
40. 6 candies per kid
41. $13 per book
42. 6 hours per day
43. $72
44. 135 minutes
45. $\frac{6}{2} = \frac{x}{18}$
46. Fancy Goods
47. Yes
48. $720
49. 62 days
50. 3 hours
51. 8 problems
52. 3 miles per hour
53. Shop B
54. $4
55. 3.5 miles per hour
56. $17
57. 120 minutes
58. 15 books per shelf
59. 5 pages per minute
60. 3 students per bench
61. 4
62. 24
63. 108
64. Yes
65. No
66. 26
67. 12
68. 9, 36
69. $120, $75
70. 28 cupcakes, 60 cupcakes
71. 48 hours
72. 0.05 miles
73. Shop A
74. Product B
75. 30 passengers per bus
76. $30
77. 30 minutes
78. 5:12
79. $12
80. $12.5
81. 490 minutes
82. C
83. 7 bananas
84. 60 problems
85. $39
86. A
87. $17.5
88. 360 bricks
89. $\frac{8}{2} = \frac{x}{18}$
90. $\frac{1}{15} = \frac{x}{90}$
91. $\frac{33}{100}$
92. $\frac{58}{10}$
93. $\frac{29}{5}$
94. $\frac{995}{1000}$
95. $\frac{203}{10000}$
96. 50%
97. 7%
98. 2.03%
99. 120%
100. 2.5%
101. $\frac{25}{100}$
102. $\frac{1}{4}$
103. $\frac{33}{25}$
104. $\frac{39}{50}$
105. $\frac{1}{200}$
106. 0.3
107. 0.0156
108. 1.5
109. 50%
110. 40%
111. 48
112. 94.5
113. $16
114. $31.5
115. 1 kg
116. 3 kg
117. 180 kg
118. 5,120 comic books
119. 801 obtained marks
120. $375
121. 168 mangoes
122. 25
123. 475
124. 43.75%
125. 50%

126. $132
127. 40%
128. 48%
129. 12.5%
130. 10%
131. 48 feet
132. 180 inches
133. 180 inches
134. 3,520 yards
135. 31,680 inches
136. $\frac{1}{4}$ foot
137. 252 inches
138. $\frac{4}{3}$ yards
139. 4.3 feet
140. 3 yards
141. 270 seconds
142. 2.5 minutes
143. 330 seconds
144. 5.42 minutes
145. 0.75 hours
146. 13 weeks
147. 64
148. 240 centimeters
149. 3 quarts
150. 4,500 grams
151. 51.2 ounces
152. B
153. 39 inches
154. A
155. 43.8 deciliters
156. $0.4375
157. Shop B
158. 4.875 gallons
159. 1 pint
160. 42 ounces
161. 3/8
162. 2/15
163. 4/27
164. 5/24
165. 2/9
166. 10/7
167. 1/3
168. 1/8
169. 6 bananas
170. 8 cup servings

171. 2 meters
172. 1/8
173. 2/7
174. 5/2
175. 5 days
176. 18 pajamas
177. 80 one-fourth pies
178. 54 two-sixth pizzas
179. 24 one-fourth broccoli stems
180. 40 half-centimeters
181. 2
182. 12
183. 7
184. 18
185. 45/4
186. 1/6
187. 7/9
188. 1/2
189. 9/20
190. ⅔
191. 8/3
192. 2/3
193. 3/2
194. 9/8
195. 23/21
196. 7/19
197. 6/5 hours or 72 minutes
198. 2 walls
199. 18 ⅔ pieces
200. 5 weeks
201. 15 m^2 wall
202. 45 m^2 wall
203. 7 pounds
204. 9 miles
205. 5 pounds of cake
206. 13 pounds of rice
207. 13 pieces
208. Sofia
209. 8
210. 3/10 hour or 18 minutes
211. 155
212. 36
213. 37

214. 138
215. 42
216. 140
217. 3620
218. 11
219. 3620
220. 1780
221. 1
222. 1420
223. 26
224. $50
225. 10 apples
226. 245 chocolates
227. 89
228. 12 pens
229. 12 tables
230. 12 months
231. $120
232. 136
233. 217 boxes
234. 100 jars
235. 15
236. 20
237. 68
238. 45, remainder 2
239. 241, remainder 2
240. 32, remainder 65
241. 84, remainder 0
242. 524, remainder 36
243. 32, remainder 77
244. 40, remainder 51
245. 23, remainder 48
246. 43, remainder 20
247. 100, remainder 12
248. 39, remainder 0
249. 312, remainder 100
250. 145, remainder 0
251. 16
252. In 4 days
253. 9 miles
254. 3 papers
255. 339 bags
256. 1,290 products
257. $150
258. 1,500 pounds of beef
259. 4 liters of petrol

260. $150	305. $0.7	349. -5
261. 0.6	306. $1,252.2	350. -3
262. 3.6	307. $6.025	351. -2
263. 55.62	308. $3.99	352. 2
264. 873.282	309. 7.5 miles per hour	353. 122
265. 0.3	310. 0.08 mile	354. -56
266. 0.54	311. 2, 3, 4, 6, 8, 24	355. -810
267. 0.6	312. 1, 2, 4, 8, 16, 32, 64	356. 10.52
268. 3.2	313. 42	357. 0
269. 60.7	314. 36	358. 109
270. 403	315. 40	359. -8.17
271. 220.003	316. 176	360. 0.32
272. 0.4	317. 12	361. Point at -2.5
273. 0.006	318. 8	362. Point at -5/2
274. 17	319. 16	363. Point at 0
275. 98.2	320. 12	364. Point at 5.5
276. 117.74	321. 240	365. Point at -3/2
277. 30 ones	322. 16	366. Point at -1
278. 2.4	323. 120	367. Point at 1
279. bc + cd	324. 12	368. Point at 2.2
280. 36.8	325. 6:40 p.m.	369. Investment income higher
281. 206.128	326. 24 miles	370. Samantha
282. 343.2284	327. 2 cm	371. Second quadrant
283. 17.0472	328. Amanda	372. B
284. 5	329. 12	373. Second quadrant
285. 70	330. 12	374. F (3, 2)
286. 290	331. zero	375. B (1, -1)
287. 44	332. positive	376. 1
288. 6.706	333. negative	377. H (-3, -2)
289. 3.44	334. negative	378. B (1, -1)
290. 3.975	335. positive	379. B (1, -1)
291. 9.635	336. -5, -3, -1, 1, 2, 7, 8, 9	380. 3
292. 5.43125	337. 8	381. C (4, 0)
293. 8.94	338. 11	382. (-6, 0), (-2, 0), (-2, 4), (-6, 4)
294. 6	339. 9	383. (-6, 4), (-2, 4), (-6, 0), (-2, 0)
295. 2.24	340. 23	384. (-4, 4), (0, 4), (-6, 0), (-2, 0)
296. 2.956	341. -10, -9, -8, -7, -6, -5, -4, -3, -2 ,-1, 0, 1, 2, 3, 4, 5	385. 8 square units
297. 1.2 pounds	342. 4	386. 6 square units
298. $15.58	343. -5	387. (1, 2), (1, 4), (4, 4), (4, 1)
299. $1,682	344. -47	388. 7.5 square units
300. 22.5 liters of milk	345. 13	
301. 18.7 miles	346. -1	
302. $3.5	347. 0	
303. $285	348. -19	
304. 64 lbs		

389. 6 square units
390. 45 students
391. 103 students
392. January
393. 75 students
394. 50 students
395. 200 students
396. June
397. June, July, August, September, October, May, April, November, March, December, February January
398. September
399. 0, 2
400. (1, 2)
401. Yes
402. Yes
403. -3, -2, 0, 5
404. -9, -7, -3, -1
405. -13, -12, 0, 15, 19
406. -2, 0, 7, 10, 15
407. -13, -12, 0, 15, 18, 22
408. -90, -10, 0, 14
409. -109, -50, -4, 12, 71
410. -49, -10, 12, 95
411. -87, -18, -15, 34
412. 5 is greater than -4
413. -10 is less than 7
414. -10 < 4 or 4 ? -10
415. B
416. C
417. 3 = 3
418. -32 < 0 or 0 > -32
419. 5 < 7 or 7 > 5
420. -32<-10 or -10>-32
421. $-\frac{1}{7}, \ -\frac{1}{20}, \ \frac{2}{5}, \ 1\frac{3}{11}, \ 4\frac{4}{7}$
422. $-\frac{7}{8}, \ -\frac{3}{10}, \ \frac{2}{5}, \ \frac{3}{7}, \ \frac{1}{2}$
423. 90°F, 50°F, 30°F, 20°F
424. New York > Montana
425. Warmer = Macomb, Cooler = New York
426. Coldest = Brooklyn, Warmest =Dallas

427. Sandi
428. New York
429. Tulip
430. 1/20
431. Gomez
432. Friday
433. Friday
434. Triangle
435. Calcium
436. Simon
437. Rectangular prism
438. 20 > 18
439. Rectangle
440. Cube
441. 1
442. 15
443. 7.15
444. 1/5
445. 209
446. 7/9
447. 79/9
448. 10
449. -23, 23, |-56|, |80|
450. -42, 4, 9, |-16|, |53|
451. 1.12
452. 410
453. 40
454. -90
455. 12.3
456. Joshua
457. Noah
458. Water
459. Joshua
460. Bird
461. $45.6
462. Jake
463. Sally
464. |-10| > |-2|
465. |-6| > |4|
466. 15
467. 38
468. Water
469. 25/9
470. $50

471. Cecilia
472. B
473. New Jersey
474. Newark, Bozeman, Macomb, New Jersey
475. Samantha, Bella, Nathalia, David
476. Checking
477. Amanda
478. Sydney
479. $219
480. Fall
481. Puppy
482. Ammonia mixture
483. Dolphin
484. New Phone
485. C
486. A
487. C, B, A
488. Boardgame
489. Shoe
490. Sweater
491. (2, -1)
492. (4, -1)
493. (0, 0), (1, 3), (4, -2), (5, 3)
494. First, y-axis; then, x-axis
495. y-axis
496. First, x-axis; then, y-axis
497. (-2, 2)
498. (2, -2)
499. (2, 2)
500. 5 units
501. 8 square units
502. No
503. No
504. No
505. (-2, -2)
506. (-2, 2)
507. (2, -2)
508. (2, 2)
509. 4 units

510. 4 units
511. Commutative
512. Commutative property of addition
513. True
514. Identity property
515. Identity property
516. Yes
517. No
518. $b \times a$
519. Yes
520. Commutative property of multiplication
521. Yes
522. Yes
523. No
524. Multiplication property of zero
525. Multiplication property of zero
526. Yes
527. Yes
528. No
529. No
530. Yes
531. Identity property of multiplication
532. Multiplicative identity property of 1
533. Yes
534. 7
535. 10,000,000
536. 76
537. $3^7 > 7^3$
538. 4^2
539. Yes
540. $7^2 \times 2^3$
541. 0.000008

542. a) False, b) True
543. 2^5
544. 8/27
545. $\frac{1}{49}$
546. $\frac{27}{125}$
547. $7^3 \times 2^2$
548. 4^5
549. 768
550. 66
551. 5
552. 2
553. 3
554. $12 - 7z$
555. $x - 2$
556. $2 \times 3 + 13 - 4$
557. $7 + 9$
558. $y + 4$
559. $\frac{y}{9}$
560. 7^9
561. $\frac{g}{4} \times 7$
562. $y - 8^4$
563. $\frac{xy}{z}$
564. $3 + 9$
565. $(a - 4) \times z$
566. $\frac{a}{2+b}$
567. $\frac{t}{2+n}$
568. $\frac{102}{p}$
569. $2 \times 10 + 3 \times y$
570. $3 + x - 1$
571. 3
572. 4
573. Yes
574. 3
575. 1
576. 2
577. $120 - 3a$
578. $\frac{4}{2} + 9$
579. $\frac{4}{x-1}$
580. $\frac{120}{3 \times 4}$

581. $- 12$
582. 12
583. 80
584. 59
585. 31
586. 88
587. 15
588. 11
589. 213
590. 23
591. $28 \ m^2$
592. $50 \ cm^3$
593. $64\pi \ m^2$
594. $16\pi \ m^2$
595. $81 \ cm^2$
596. $21 \ m^2$
597. $104°$ Fahrenheit
598. 2
599. 3%
600. 10
601. 14
602. 7.71
603. $\frac{19}{90}$
604. 53
605. 50
606. $\frac{3}{2}$
607. 15
608. 14
609.
$\$200 - \frac{1}{2} \times (\$200) - 3 \times (\$4)$
$= \$88$
610. $60°$ *Celsius*
611. Increase the value of x.
612. Decrease the value of a.
613. Nothing happens
614. 900
615. 81.5
616. 113
617. $\frac{796}{3}$
618. $4 \times 9^2 > 12^2 + 3^3$

619. -1
620. 14.6
621. $5xy$
622. $50pq$
623. $20xy$
624. $14xy$
625. $27xy$
626. 5×4×x×y
627. 11×7×s×t
628. 5×7×a×b
629. (i) Commutative property of addition
(ii) Add
630. (i) Associative property of addition
(ii) Add
631. (i) Distributive property
(ii) Multiply
632. (i) Associative property of multiplication
(ii) multiply
633. $16x + 20y$
634. $5x + 10y$
635. $4(4x + 5y)$
636. $5(x + 2y)$
637. $20x + 9$
638. $12x + 6$
639. $12x$
640. $12x + 2y$
641. $5p$
642. $20p$
643. $8x + 16z$
644. $3x + 4z$
645. $0.2x + 0.4z$
646. $2(27x + y)$
647. $5(x + 4y + 10z)$
648. $9(5 + 6y)$
649. Yes

650. True
651. Yes
652. (a)+(ii), (b)+(iii), (c)+(i)
653. (a)+(ii), (b)+(i), (c)+(iii)
654. (a)+(iii), (b)+(ii), (c)+(i)
655. No
656. $-28 + 92x$
657. $2x$
658. $9p$
659. 10x+8y
660. 4x+4y
661. 7x+5
662. $-6p$
663. $3x + 9z$
664. $3(4 - 2x - z)$
665. 24(1+2)
666. 38
667. $5l$
668. Yes
669. False
670. True
671. True
672. Yes
673. No
674. No
675. Yes
676. Yes
677. Yes
678. $\frac{39}{10}$
679. 0.8
680. No
681. 4
682. 4
683. $\frac{1}{4}$ *th* of the bread
684. $2x - 4 = 10$
   Step1: $2x - 4 + 4 = 10 + 4$
   Step2: $2x = 14$
   Step3: $\frac{2x}{2} = \frac{14}{2}$
Finally, $x = 7$

685. 15
686. -151
687. 87
688. 4.6
689. -1
690. 1, 2, 3, 4
691. -1, -2, -3, -4
692. 1, 2, 5, 6
693. 6, 7, 8, 9
694. Yes
695. Yes
696. Yes
697. Yes
698. 7, 8, 9
699. No
700. 2, 4, 6, 8
701. 9 - a apples
702. $2x
703. $2x
704. $y \geq 3$
705. $x + y + z$
706. $x + 2$ *cm*
707. $x - 3$ *cm*
708. $x - 2$ pieces
709. $x + 20 = 30$
710. 10
711. $5a = 25$
712. 5
713. $20 + x$ blueberries
714. $20 + x$
715. 67 stamps
716. $\frac{36}{N} = 4$
717. 9
718. $4y = 12$
719. 3
720. $8w = 96$
721. 12
722. 36
723. 2

724. 105
725. 6
726. $\frac{2}{7}$
727. 18
728. 2
729. 4
730. 23
731. 6
732. 74
733. 57
734. 7
735. $\frac{2}{3}$
736. $\frac{315}{2}$
737. $\frac{229}{30}$
738. 72
739. 3.6
740. 6
741. 5
742. 14
743. $\frac{1}{2}$
744. $x + 2 = 9$
745. $\frac{25}{w} = 2.5$
746. $x + 5 = 57$
747. $\frac{w}{2} = 7$
748. 14
749. 3
750. 64
751. $x > 10$
752. $y > 2$
753. $y > 10$
754. $y \geq 2$
755. $b = a$
756. $h > 5$
757. 455
758. $\frac{18}{7}$
759. Step 1
760. 7
761. $x > 0$
762. $x \geq 11$

763. $x \leq 48$
764. Yes
765. $y > -1$
766. $g \leq 2$
767. $r \geq 9$
768. See page 258
769. See page 258
770. $x \leq -\frac{35}{3}$
771. b is the independent variable and a is the dependent variable.
772. y is the independent variable and x is the dependent variable.
773. t is the independent variable and c is the dependent variable.
774. c is the independent variable and t is the dependent variable.
775. c is the independent variable and t is the dependent variable.
776. y is the independent variable and x is the dependent variable.
777. 2
778. G is the independent variable and S is the dependent variable.
779. 7
780. G is the independent variable and S is the dependent variable.

781. 7
782. 2
783. 20 years old
784. $775
785. 30
786. $4,600
787. 500
788. 50
789. See page 259
790. See page 259
791. $25
792. $5
793. (a) 9, (b) 7
794. See page 260
795. (b) 12, (c) 23
796. See page 260
797. $y = x + 1$
798. $w = 3z$
799. 25
$g = 2f + 1$

800. 48
$g = 3f$

801. $216m^2$

802. $10m^2$

803. $35\ cm^2$

804. $5\ m$

805. 30 square inches

806. 21 square inches

807. $16\ m$

808. 16 m

809. $42\ yd^2$

810. $81\ mm^2$

811. $20\ ft^2$

812. $42\ ft^2$

813. 49 cm^2	837. $552	860. 112 square yd
814. 56 m^2	838. 144 cm^3	861. See page 263
815. 50 mm^2	839. 96 mm^3	862. See page 263
816. 548 ft^2	840. 5,776 boxes	863. 150 square cm
817. 30 in^2	841. (2, 6) or (4, 6)	864. 1.5 square mm
818. 72 in^2	842. (4, 1)	865. 1.48 square units
819. 5 in.	843. 2.5 square units	866. 9+9+9+9+9+9
820. 12 m	844. (1.5, 1), (3.5, 1), (1.5, 3.5)	867. 42+36+36+36+42
821. 120 m^3		868. 85+120+408+85+120+408
822. 729 m^3	845. (2.25, 3.5), (1.5, 2), (4.5, 2), (3.75, 3.5)	
823. 180 mm^3	846. 6 square units	869. 100+30+30+30+30
824. 8 ft	847. 12 square units	870. 15.36+27+27+27
825. 7 cm^3	848. See page 261	871. No
826. 8 yd	849. See page 261	872. No
827. 15, 625 mm^3	850. y > x	873. No
		874. Yes
828. 396 in^2	851. 294 square mm.	875. No
829. $\frac{1729}{8}$ cm^3	852. 222 square units	876. No
		877. Yes
830. 72 cm^3	853. 8.5 square meters	878. No
831. 576π cm^3	854. 128.8 square cm	879. No
832. 5π m^3	855. 507.84 square ft	880. Yes
		881. 13
833. 46 $cubic\ meters$	856. Cube	882. 7
834. 126 m^3	857. Rectangular prism	883. 4
		884. 7
835. 216 mm^3 $iron$	858. See page 262	885. 3
		886. 9
836. 97.4 cm^3	859. See page 262	887. 5.5
		888. 8
		889. symmetric
		890. 6.5

891. 4.14	930. 16 books	955. 5
892. 3.5		
893. 3	931. 50	956. 11
894. 35.71		
895. 34	932. 2020	957. 7
896. 15.6		
897. 11 employees	933. Yes.	958. 8
898. 12.33		
899. 13 employees	934. No.	959. 11
900. 92		
901. 3 candies	935. Yes.	960. 160
902. 6 candies		
903. 1 problem	936. No.	961. 15.5
904. 7 problems		
905. 102°F	937. 130 flowers	962. 18
906. 100°F		
907. 30-34 kg	938. 90 buildings	963. 33
908. 6 students	939. 70	964. 18
909. 25 students		
910. 18 students	940. 10 candies	965. 7
911. 6		
912. 11°	941. 2	966. 8
913. 10 apples		
914. 23.8	942. 5	967. 9
915. 4	943. 2, 11	968. 6
916. Group 4		
917. 5	944. 7	969. 8
918. 7		
919. 7	945. 6	970. 7, 24
920. 11	946. 59	971. 75 people
921. 1990		
922. $55	947. 8	972. CBS
	948. 8	973. 320 customers
923. 100 detective books	949. 45	974. 50
924. 3 students	950. 8	975. 150
925. 5 families	951. 78	976. 400
926. 46 children	952. 18	977. 30
927. 4 candies	953. 6,602	978. 25
928. 2 students	954. 13	979. The mean will decrease.
929. 2 students		

980. The mean will increase.

981. 3

982. 15

983. 1.5

984. 45

985. 32

986. 77

987. 53

988. 25%

989. 75%

990. False

991. The mean decreased.

992. The median decreased.

993. The mean will increase.

994. The mean will decrease.

995. The mean will decrease

996. No, the median will decrease.

997. The mean will decrease from $2,100 to $1,950.

998. The median will decrease from $2,000 to $1,900.

999. The mean will decrease.

1000. Yes, the median will remain the same.

768.

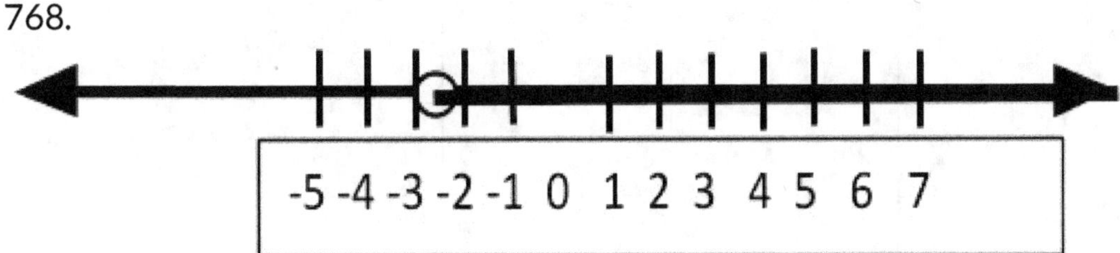

769.

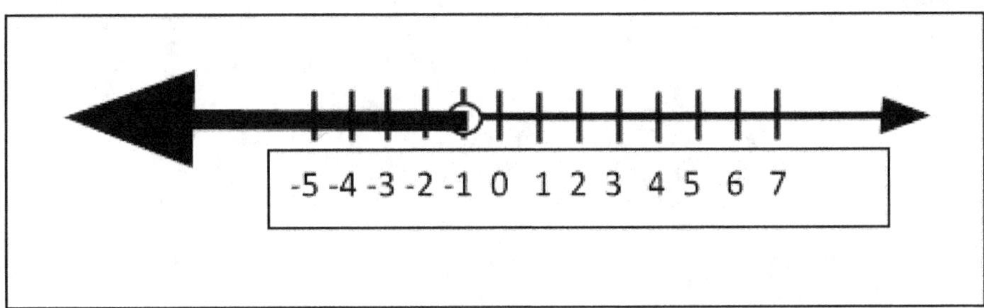

789.

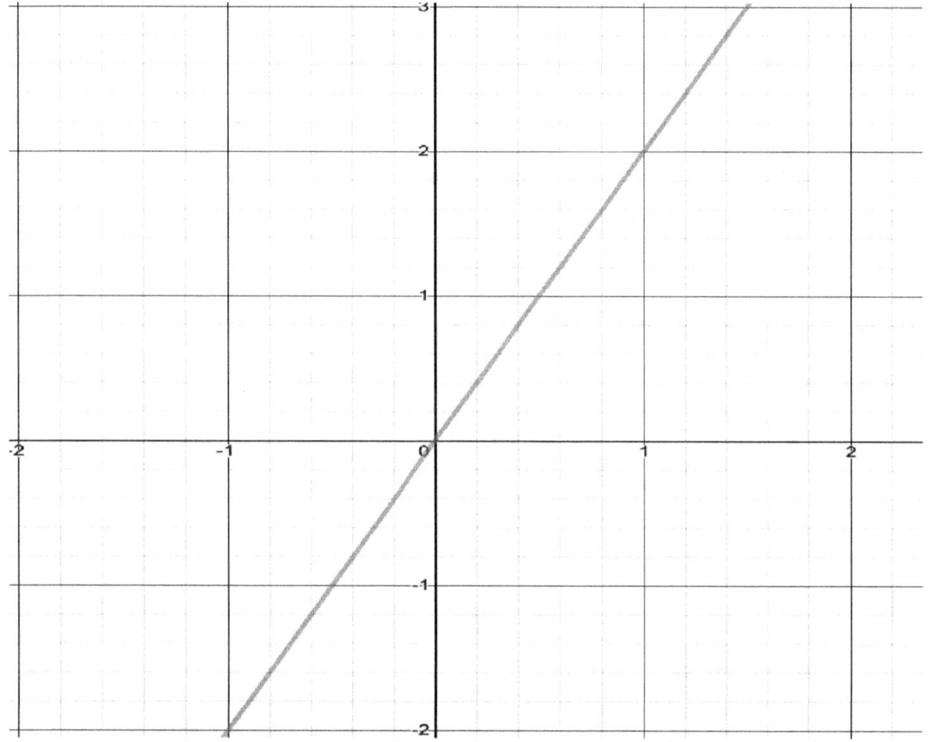

790.

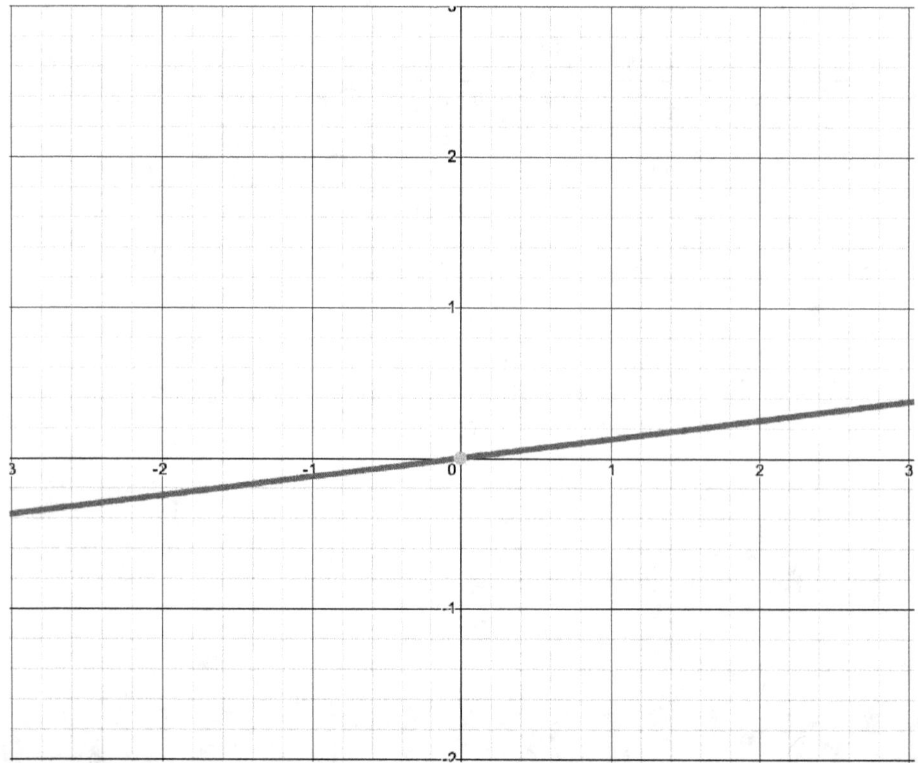

794.

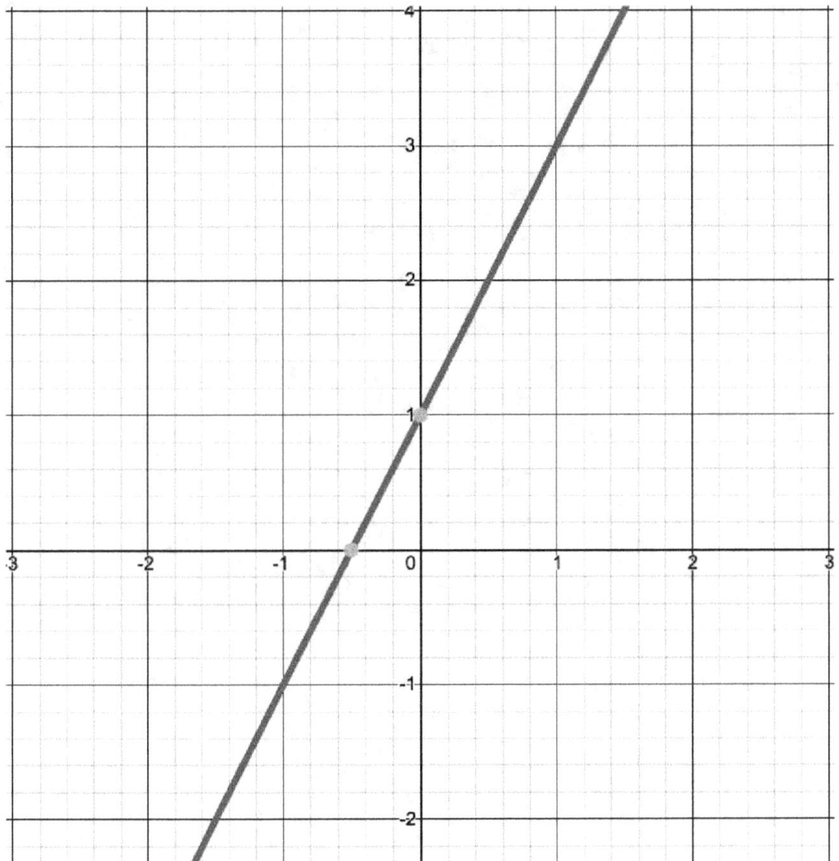

796.

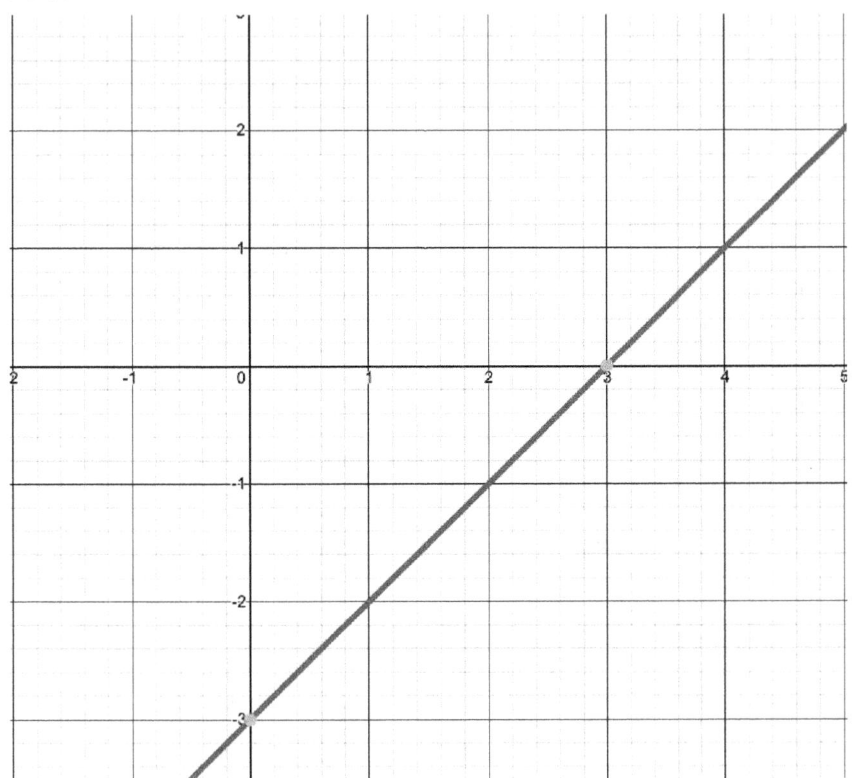

848.

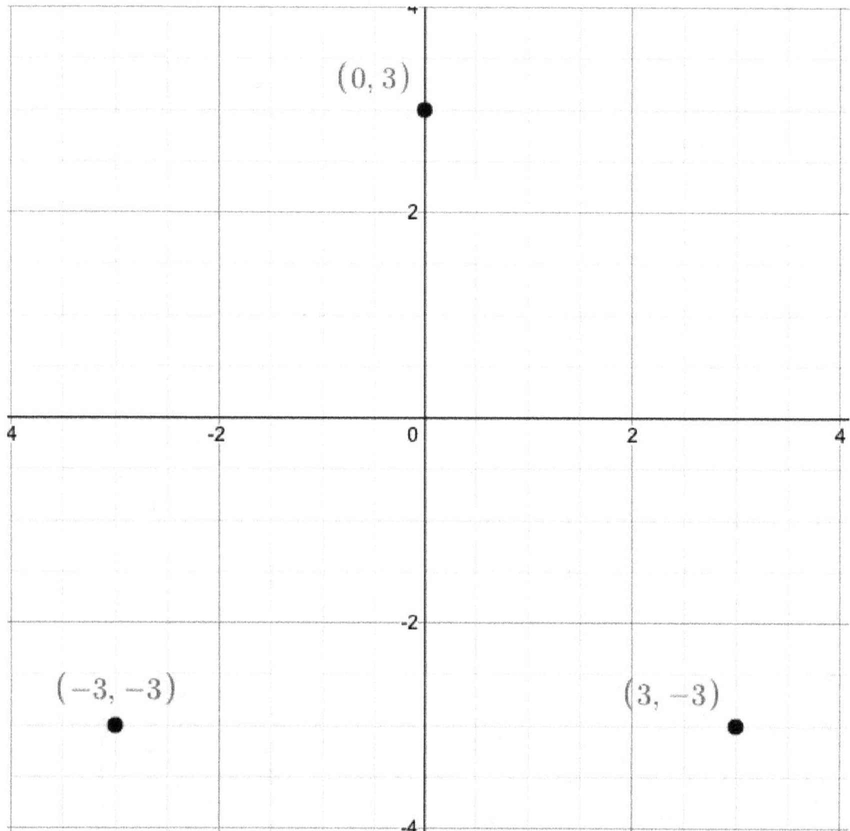

849.

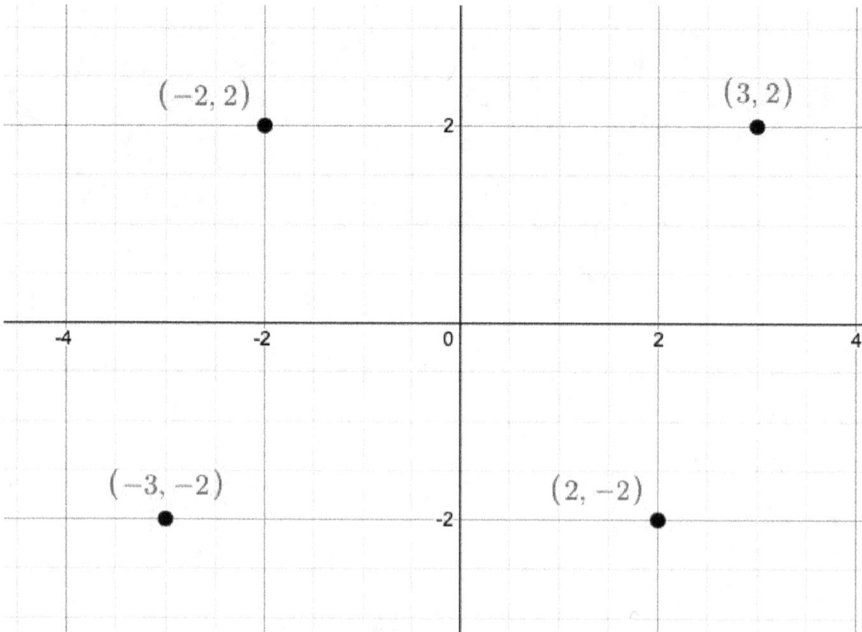

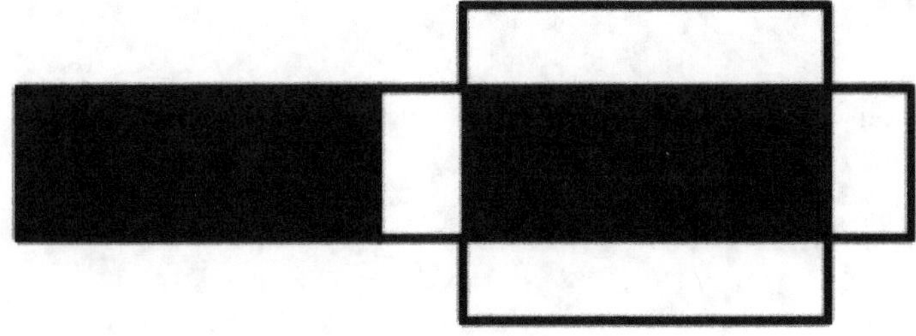

859. Triangular prism

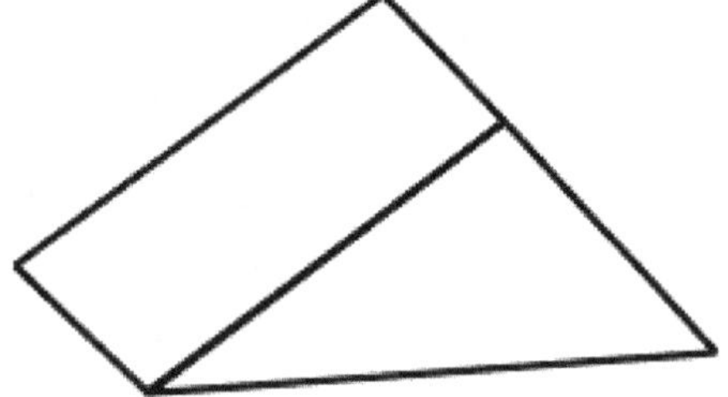

861. Triangular pyramid

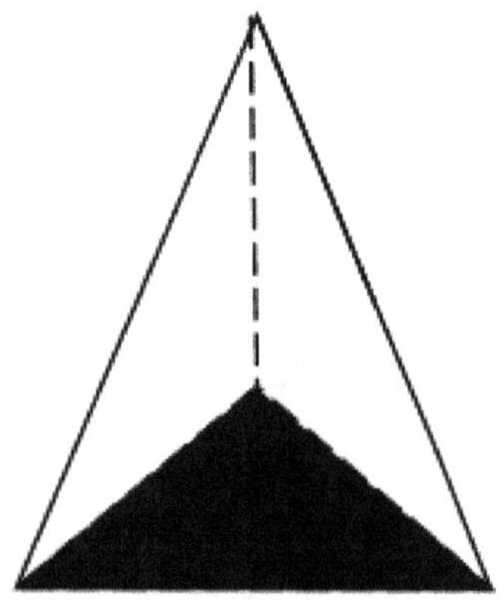

862.